ISSN 0081-4539

2012

THE STATE OF FOOD AND AGRICULTURE

FOOD AND AGRICULTURE ORGANIZATION OF THE UNITED NATIONS
Rome, 2012

ISBN 978-92-5-107317-9

Contents

TABLES

BOXES

FIGURES

Foreword

Heads of State and Government and high-level representatives from countries and organizations gathered at the United Nations Conference on Sustainable Development (Rio+20) in June 2012 to declare a common commitment to ensuring the promotion of an economically, socially and environmentally sustainable future for our planet and for present and future generations. Agriculture and hunger eradication have also taken their rightful place as one of the top priorities on the international agenda. During the Rio+20 gathering, the United Nations Secretary-General Ban Ki-moon announced The Zero-Hunger Challenge, calling for an end to world hunger. This edition of *The State of Food and Agriculture, "Investing in agriculture for a better future"*, makes the case that increasing the levels and the quality of investment in agriculture is central to achieving these goals. It also argues that we need to change the way we invest in agriculture.

Investing in agriculture is one of the most effective ways of promoting agricultural productivity, reducing poverty and enhancing environmental sustainability. Making the transition to sustainable agriculture will not be possible without significant new investment to protect and enhance the efficiency of natural resource use and to reduce waste at all stages of production, processing and consumption. Yet levels of private and public investment per worker in agriculture are stagnant or falling in the regions where rural poverty and hunger are most severe. What's more, too often government spending on agriculture does not yield the highest returns in terms of agricultural productivity, poverty reduction and sustainability.

There is no doubt that more public resources are needed for agriculture. However, rather than just advocating more government and donor funding, this report calls for a new investment strategy that puts agricultural producers at its centre and focuses public resources at all levels on the provision of public goods and the creation of an enabling environment for investment by farmers. It calls upon governments at all levels and their development partners to channel both public and private investment towards activities that yield higher returns for society. All agricultural investors and rural businesses need good governance, macro-economic stability, rural infrastructure, secure property rights and effective market institutions in order to mobilize the resources and assume the significant risks that investing in agriculture entails.

Investors at the smallest and largest ends of the spectrum require special attention: smallholders need support in overcoming the constraints they face in saving and accumulating assets and in coping with the uncertainty and risk that are intrinsic to farming. Cooperatives and other types of producer organizations can help smallholders confront some of these challenges, and social safety nets can allow the poorest farm households to escape poverty traps that prevent them from building productive assets. Large-scale investment may offer opportunities in terms of increased production, export earnings, employment and technology transfer, but they require good governance to protect the rights of local communities and to avoid natural resource degradation.

Investing in agriculture for a better future involves more than simply accumulating physical capital in the sector, although this is part of the challenge; it requires building the institutions and human capacity that will allow the agriculture sector to contribute to a sustainable future. It is my hope that this report will stimulate the global community to put agricultural producers at the centre of their investment strategies for the sector. Only by catalysing investment by farmers and directing public investment appropriately can we achieve a world in which everyone is well nourished and natural resources are used sustainably.

José Graziano da Silva
FAO DIRECTOR-GENERAL

Acknowledgements

The State of Food and Agriculture 2012 was prepared by members of the Agricultural Development Economics Division (ESA) of FAO under the overall leadership of Kostas Stamoulis, Director; Keith Wiebe, Principal Officer; and Terri Raney, Senior Economist. Technical guidance was provided by Pietro Gennari, Director of the Statistics Division (ESS); David Hallam, Director of the Trade and Markets Division (EST); Richard China, Director of the Policy and Programme Development Support Division (TCS); and Charles Riemenschneider, Director of the Investment Centre Division (TCI). Additional guidance was provided by Ann Tutwiler, Deputy Director-General (Knowledge); Marcela Villarreal, Director, and Eve Crowley, Principal Officer, of the Gender, Equity and Rural Employment Division (ESW); Josef Schmidhuber, Principal Officer (ESS) and Boubaker Benbelhassen, Principal Officer (EST).

Part I was prepared by a research and writing team led by Jakob Skoet and including Gustavo Anríquez, Brian Carisma, André Croppenstedt, Sarah Lowder, Ira Matuschke, Terri Raney and Ellen Wielezynski, all from ESA. Additional inputs from ESA were provided by Jean Balié, Jesús Barreiro Hurlé, Benjamin Davis, Paulo Dias, Lauren Edwards, Panagyotis Karfakis, Marco Knowles, Leslie Lipper, George Rapsomanikis, Cameron Short, Julian Thomas, Antonio Vezzani and Tiantian Zha. Other contributors included Pascal Liu (EST); Maria Adelaide D'Arcangelo, Ana Paula de la O Campos, Denis Herbel, Marta Osorio, Nora Ourabah Haddad and Clara Park (ESW); Masahiro Miyazako and Saifullah Syed (TCS); Calvin Miller Rural Infrastructure and Agro-Industries Division; Astrid Agostini, Tommaso Alacevich, Eugenia Serova, Garry Smith and Benoist Veillerette (TCI); David Palmer (Climate Energy and Tenure Division); and Ciro Fiorillo (FAO Country Office in Bangladesh).

Several international datasets were revised and updated for this report. Much of the analysis in the report would not have been possible without the efforts of Dominic Ballayan, Carola Fabi, Ilio Fornasero, Amanda Gordon, Erdgin Mane, Robert Mayo and Pratap Narain (all of ESS), who generated the data on agricultural capital stock and other statistical indicators and provided advice on other data sources. The team are also grateful to staff members of IFPRI, including Bingxin Yu and Sangeetha Malaiyandi for data on government expenditures on agriculture from the SPEED database; Sam Benin for data on African government expenditures from the ReSAKSS database; and Nienke Beintema, Michael Rahija and Gert-Jan Stads for data and analysis on agricultural research and development from the Agricultural Science and Technology Indicators (ASTI) project. Masataka Fujita of UNCTAD was instrumental in making available the data on Foreign Direct Investment; and Yasmin Ahmad of OECD answered queries on the data on Official Development Assistance.

Background papers and additional inputs were prepared by Kym Anderson (University of Adelaide); Michel Benoit-Cattin (CIRAD MOISA Montpellier); Christian Böber (University of Hohenheim); Nadia Cuffaro (Universitá degli Studi di Cassino); Stefan Dercon (University of Oxford); Mahendra Dev (Indira Gandhi Institute of Development Research); Shenggen Fan, Linden McBride, Tewodaj Mogues and Bingxin Yu (all from IFPRI); Keith Fuglie (Economic Research Service, United States Department of Agriculture); Ron Kopicki (former World Bank); David Lee (Cornell University); Carly Petracco (European Bank for Reconstruction and Development); and Bettina Prato (IFAD).

The report benefited from two technical workshops, with the participation of Derek Byerlee, Rita Butzer (University of Chicago), Cesar Falconí (Inter-American Development Bank), Madhur Gautam (World Bank), Donald Larson (World Bank), Ellen McCullough (Bill and Melinda Gates Foundation), Tewodaj Mogues (IFPRI), Bettina Prato (IFAD), Philippe Rémy (IFAD), Carlos Seré (IFAD), Gert-Jan Stads (IFPRI), Alberto Valdés (Universidad Católica de Chile), Bingxin Yu (IFPRI), Linxiu Zhang (Center for Chinese Agricultural Policy,

Chinese Academy of Sciences). The writing team is most grateful to the workshop participants and many other internal and external reviewers of various drafts of the manuscript.

Financial support was provided by IFAD and the Government of Japan (project on "Support to Study on Appropriate Policy Measures to Increase Investment in Agriculture and Stimulate Food Production") for data acquisition and analysis, preparation of background papers and organization of workshops. The team gratefully acknowledges this support.

Part II of the report was prepared by Merritt Cluff and Holger Matthey (EST) under the guidance of Jakob Skoet.

Part III of the report was prepared by Sarah Lowder with assistance from Brian Carisma (both from ESA). It was reviewed by Aparajita Bijapurkar (ESA).

Michelle Kendrick (Economic and Social Department of FAO) was responsible for English editorial work and project management. Paola Di Santo and Liliana Maldonado provided administrative support throughout the process. Annelies Deuss (Carnegie Mellon University) reviewed the final draft of the report. Translations and printing services were provided by the FAO Meeting Programming and Documentation Service. Graphic, layout and proofing services were provided by Omar Bolbol, Flora Dicarlo and Green Ink.

Abbreviations and acronyms

AOI	Agricultural Orientation Index
ASTI	Agricultural Science and Technology Indicators
CAADP	Comprehensive Africa Agriculture Development Programme
CFS	Committee on World Food Security
EU	European Union
FDI	foreign direct investment
FPI	Food Price Index (FAO)
G20	Group of Twenty Finance Ministers and Central Bank Governors
G8	Group of Eight
GAFSP	Global Agriculture and Food Security Program
GDP	gross domestic product
GEF	Global Environment Facility
IFAD	International Fund for Agricultural Development
IFPRI	International Food Policy Research Institute
IMF	International Monetary Fund
MAFAP	Monitoring African Food and Agricultural Policies
MDG	Millennium Development Goal
NEPAD	New Partnership for Africa's Development
NFP	National Food Policy
NGO	non-governmental organization
NRP	nominal rate of protection
ODA	official development assistance
OECD	Organisation for Economic Co-operation and Development
PRAI	*Principles for Responsible Agricultural Investment that Respects Rights, Livelihoods and Resources*
PSNP	Productive Safety Net Programme
PSTA II	Strategic Plan for Agriculture Transformation II
R&D	research and development
RRA	relative rate of assistance
ReSAKSS	Regional Strategic Analysis and Knowledge Support System
SPEED	Statistics of Public Expenditure for Economic Development
TFP	total factor productivity
UNCTAD	United Nations Conference on Trade and Development
USAID	United States Agency for International Development
VGGT	*Voluntary Guidelines on the Responsible Governance of Tenure of Land, Fisheries and Forests in the Context of National Food Security*

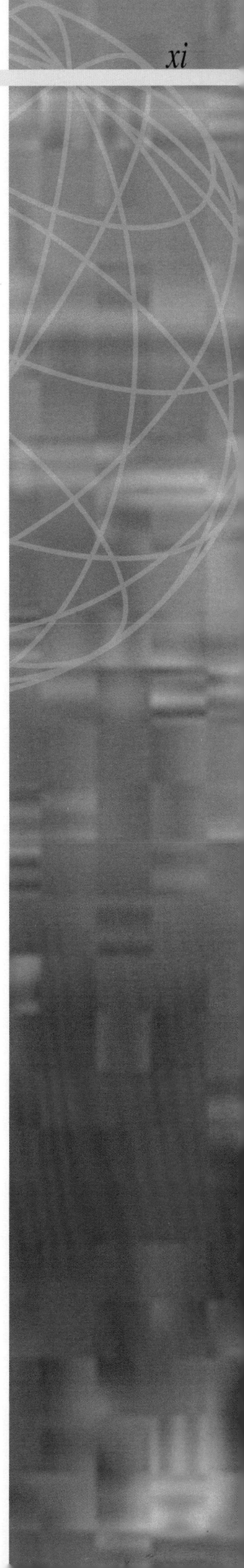

Executive summary

The State of Food and Agriculture 2012: Investing in agriculture for a better future shows that farmers are the largest investors in developing country agriculture and argues, therefore, that farmers and their investment decisions must be central to any strategy aimed at improving agricultural investment. The report also presents evidence showing how public resources can be used more effectively to catalyse private investment, especially by farmers themselves, and to channel public and private resources towards more socially beneficial outcomes. The focus of this report is on the accumulation of capital by farmers in agriculture and the investments made by governments to facilitate this accumulation.

Agricultural investment is essential to promoting agricultural growth, reducing poverty and hunger, and promoting environmental sustainability. The regions of the world where hunger and extreme poverty are most widespread today – South Asia and sub-Saharan Africa – have seen stagnant or declining rates of investment per worker in agriculture for three decades. Recent evidence shows signs of improvement, but eradicating hunger in these and other regions, and achieving this sustainably, will require substantial increases in the level of farm investment in agriculture and dramatic improvements in both the level and quality of government investment in the sector.

Farmers must be central to any investment strategy

This report presents the most comprehensive data that has been prepared to date on the relative sizes of investment and expenditure flows by farmers, governments, donors and private foreign investors in low- and middle-income countries. Public and private investors spend their resources on different things and for different reasons, and it is not always easy to distinguish between investment and expenditures. In simple terms, investment involves accumulating assets that generate increased income or other benefits in the future, while expenditures also involve current expenses and transfer payments that are not normally considered investment.

Despite these conceptual and empirical limitations, the best available data show that farmers in low- and middle-income countries invest more than four times as much in capital stock on their own farms each year as their governments invest in the agriculture sector. What's more, farmers' investment dwarfs expenditures on agriculture by international donors and private foreign investors. The overwhelming dominance of farmers' own investment means that they must be central to any strategy aimed at increasing the quantity and effectiveness of agricultural investment.

A conducive investment climate is essential for agriculture

Farmers' investment decisions are directly influenced by the investment climate within which they operate. While many farmers invest even in unsupportive investment climates (because they may have few alternatives), a large body of evidence discussed in this report shows that farmers invest more in the presence of a conducive investment climate and that their investment is more likely to have socially and economically beneficial outcomes.

The existence or absence of a conducive investment climate depends on markets and governments. Markets generate price incentives that signal to farmers and other private entrepreneurs when and where opportunities exist for making profitable investments. Governments are responsible for creating the legal, policy and institutional environment that enables private investors to respond to market opportunities in socially responsible ways. In the absence of an enabling environment and adequate market incentives, farmers will not invest adequately in agriculture and their investment may not yield socially optimal results. Indeed, building and maintaining the enabling environment for private investment is itself one of the most important investments that can be made by the public sector.

The elements of a good general investment climate are well known, and many of the same factors are equally or more important in the enabling environment for agriculture: good governance, macroeconomic stability, transparent and stable trade policies, effective market institutions and respect for property rights. Governments also influence the market incentives for investment in agriculture relative to other sectors through support or taxation of the agriculture sector, exchange rates and trade policies, so care must be taken to ensure equitable treatment of agriculture. Ensuring an appropriate framework for investment in agriculture also requires the incorporation of environmental costs and benefits into the economic incentives facing investors in agriculture and the establishment of mechanisms facilitating the transition to sustainable production systems.

Governments can help smallholders overcome challenges to investment

Farmers in many low- and middle-income countries face an unconducive environment and weak incentives to invest in agriculture. Smallholders often face specific constraints, including extreme poverty, weak property rights, poor access to markets and financial services, vulnerability to shocks and limited ability to tolerate risk. Ensuring a level playing field between smallholders and larger investors is important for reasons of both equity and economic efficiency. This is particularly the case for women engaged in agriculture, who often encounter even more severe constraints. Effective and inclusive producer organizations can allow smallholders to overcome some of the constraints relating to access to markets, natural resources and financial services. Social transfers and safety net schemes can also play a role as policy instruments to allow the poorest smallholders to expand their asset base. These can be instrumental in overcoming two of the most severe constraints faced by poor smallholders: lack of own savings and access to credit and lack of insurance against risks. Such mechanisms can allow poor smallholders and rural households to build assets and overcome poverty traps, but their choice of assets (human, physical, natural or financial capital) and activities (farming or non-farm activities) will depend on the overall incentive structure as well as the households' individual circumstances.

Large-scale private investment offers opportunities but requires governance

The increasing international flow of funds directed towards large-scale land acquisitions by private companies, investment funds and sovereign wealth funds has been receiving significant attention. The limited scale of such investment means it is likely to have only a marginal impact in terms of global agricultural production. However, the potential impact at the local level as well as the potential for future growth has led to concerns about possible negative social and environmental impacts, especially in low-income countries, which often have less capacity to establish and implement a regulatory framework to address these issues.

Large-scale investment may offer opportunities to increase production and export earnings, generate employment and promote technology transfer, but can involve risks in terms of overriding the rights of existing land users and generating negative environmental impacts. A clear challenge is to improve the capacity of governments and local communities to negotiate contracts that respect the rights of local communities as well as their ability to monitor and enforce them. Instruments such as the *Principles for Responsible Agricultural Investment that Respects Rights, Livelihoods and Resources* and the *Voluntary Guidelines for the Responsible Governance of Tenure of Land, Fisheries and Forests in the Context of National Food Security* offer a framework in this regard. Alternative and more inclusive business models for large-scale investors that offer opportunities for greater direct involvement of local farmers in agricultural value chains should be promoted.

Investing in public goods yields high returns in agricultural growth and poverty reduction

The provision of public goods is a fundamental part of the enabling environment for agricultural investment.

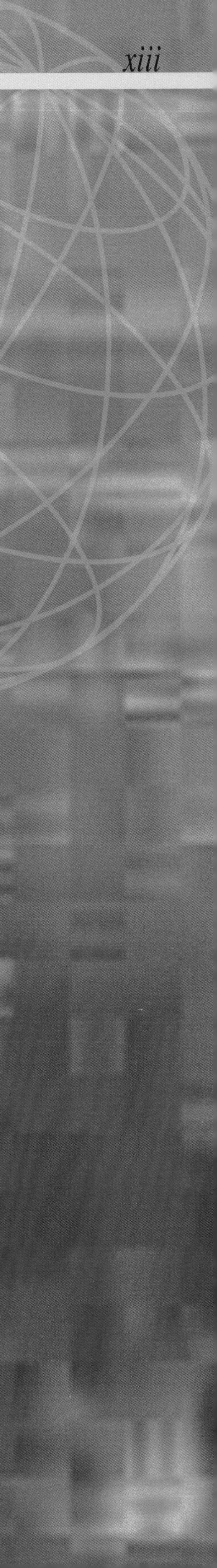

Evidence from many countries over five decades shows that public investment in agricultural research and development (R&D), education and rural infrastructure yields much higher returns than other expenditures such as input subsidies. Investing in public goods for agriculture yields strong returns in terms of both agricultural productivity and poverty reduction, indicating that these are usually compatible, not competing, goals. Investments in public goods in rural areas are also likely to be complementary in nature; investments in education and rural infrastructure tend to enhance agricultural investment and are often ranked among the top sources of agricultural growth and overall economic growth in rural areas. The relative impact of alternative investments varies by country, so priorities for investment must be locally determined, but the returns to investment in public goods in rural areas are mutually reinforcing.

Improving the performance of public expenditures

In spite of the extensive body of evidence documenting high economic and social returns on investment in public goods that directly and indirectly support agriculture, government budget allocations do not always reflect this priority, and actual spending does not always reflect budget allocations. A number of political economy factors are to blame, including collective action by powerful interest groups, difficulties in attributing responsibility for successful investments that have long lead times and diffuse benefits (as do many agricultural and rural public goods), poor governance and corruption. Strengthening rural institutions and promoting transparency in decision-making can improve the performance of governments and donors in ensuring that scarce public resources are allocated to the most socially beneficial outcomes. Many governments are making efforts to improve the planning, targeting and efficiency of their expenditures, including more transparent and inclusive budget processes. Much more needs to be done to encourage these efforts.

Key messages of the report

- **Investing in agriculture is one of the most effective strategies for reducing poverty and hunger and promoting sustainability.** The regions where agricultural capital per worker and public agricultural spending per worker have stagnated or fallen during the past three decades are also the epicentres of poverty and hunger in the world today. Demand growth for agricultural products over the coming decades will put increasing pressure on the natural resource base, which in many developing regions is already severely degraded. Investment is needed for conservation of natural resources and the transition to sustainable production. Eradicating hunger sustainably will require a significant increase in agricultural investment and, more importantly, it will require improving the quality of investment.
- **Farmers are by far the largest source of investment in agriculture.** In spite of recent attention to foreign direct investment and official development assistance, and in spite of weak enabling environments faced by many farmers, on-farm investment by farmers themselves dwarfs these sources of investment and also significantly exceeds investments by governments. On-farm investment in agricultural capital stock is more than three times as large as other sources of investment combined.
- **Farmers must be central to any strategy for increasing investment in the sector, but they will not invest adequately unless the public sector fosters an appropriate climate for agricultural investment.** The basic requirements are well known, but too often ignored. Poor governance, absence of rule of law, high levels of corruption, insecure property rights, arbitrary trade rules, taxation of agriculture relative to other sectors, failure to provide adequate infrastructure and public services in rural areas and waste of scarce public resources all increase the costs and risks associated with agriculture and drastically reduce incentives for investment in the sector.

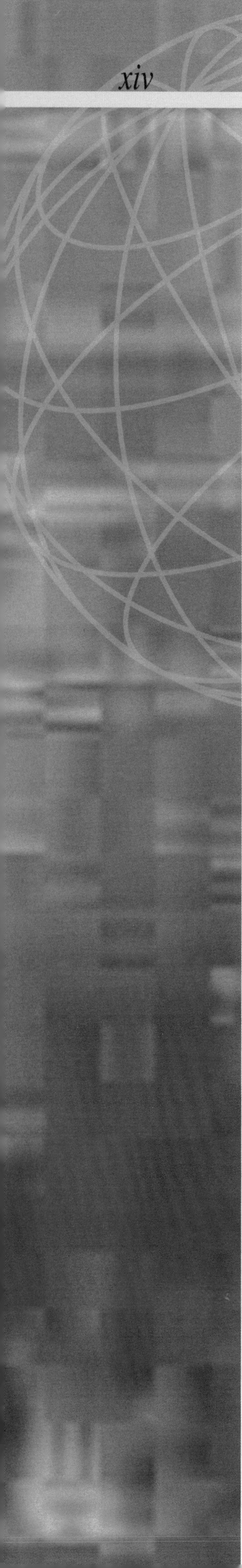

Governments must invest in building the institutions and human capacity necessary to support an enabling environment for agricultural investment.

- **A favourable investment climate is indispensable for investment in agriculture, but it is not sufficient to allow many smallholders to invest and to ensure that large-scale investment meets socially desirable goals.**
  - **Governments and donors have a special responsibility to help smallholders overcome barriers to savings and investment.** Smallholders often face particularly severe constraints to investing in agriculture because they operate so close to the margins of survival that they are unable to save or to tolerate additional risk. They need more secure property rights and better rural infrastructure and public services. Stronger producer organizations such as cooperatives would help them manage risks and achieve economies of scale in accessing markets. Social safety nets and transfer payments may help them accumulate and retain assets, either in agriculture or in other activities of their choice.
  - **Governments, international organizations, civil society and corporate investors must ensure that large-scale investments in agriculture are socially beneficial and environmentally sustainable.** Large-scale investments, including by foreign corporations and sovereign investors, may offer opportunities for employment and technology transfer in agriculture but may also pose risks to the livelihoods of local populations, especially in cases of unclear property rights. Governance of these investments must be improved by promoting transparency, accountability and inclusive partnership models that do not involve transfer of land and that allow local populations to benefit.
- **Governments and donors need to channel their limited public funds towards the provision of essential public goods with high economic and social returns.** Public investment priorities will vary by location and over time; but evidence is clear that some types of spending are better than others. Investment in public goods such as productivity-enhancing agricultural research, rural roads and education have consistently higher payoffs for society than spending on fertilizer subsidies, for example, which are often captured by rural elites and distributed in ways that undermine private input suppliers. Such subsidies may be politically popular, but they are not usually the best use of public funds. By focusing on public goods, including sustainable natural resource management, governments can enhance the impact of public expenditures in terms of both agricultural growth and poverty reduction. Governments must invest in building the institutions and human capacity necessary to support an enabling environment for agricultural investment.

Part I

INVESTING IN AGRICULTURE FOR A BETTER FUTURE

Part I

1. Introduction

Recent food crises and growing concerns about global climate change have placed agriculture on top of the international agenda. Governments, international organizations, and civil society groups gathered at the Group of Eight (G8), the Group of Twenty Finance Ministers and Central Bank Governors (G20) and Rio+20 summits in 2012 have recognized a convergence between the dual goals of eradicating hunger and making agriculture sustainable. Achieving these goals will require a significant increase in agricultural investment but, more importantly, it will require improving the quality of this investment.

FAO has long advocated investing in agriculture. The first edition of *The State of Food and Agriculture*, published in 1947, identified the need for more investment in agriculture to produce food for deficit regions, and the 1949 edition reported financial targets for levels of investment required to rebuild agriculture after the Second World War (FAO, 1947; FAO, 1949). These and many subsequent reports focused on the role of governments in planning and directing the investment requirements for agriculture, with little attention to the role of farmers themselves.

The international financial crisis, which is affecting governments and donors around the world, means that now, more than ever, public resources alone cannot meet the investment needs for agriculture. Governments and donors play a crucial role in catalysing, channelling and governing agricultural investment, but private investors – primarily farmers themselves – must be central to any investment strategy for agriculture.

This edition of *The State of Food and Agriculture* reviews the economic and social rationale for agricultural investment, examines the causes of underinvestment in agriculture and presents evidence showing how public resources can be used more effectively. The focus of this report is on the accumulation of capital by farmers in agriculture and the investments made by governments to facilitate this accumulation. *Investing in agriculture for a better future* can help achieve a world in which everyone is well nourished and natural resources are used sustainably.

Who invests in agriculture?

Investors in agriculture can be categorized as public or private and foreign or domestic.[1] The majority of private domestic investors are farmers and they are by far the largest source of investment in agriculture in low- and middle-income countries. Domestic public investors, primarily national governments, are the next largest source of investment in agriculture, followed distantly by foreign public investors such as development partners and by foreign private investors, such as corporations. These investors – public and private, domestic and foreign – invest in different things and for different reasons. Their investments are often complementary, sometimes overlapping, and are generally

[1] In this report, "agriculture" refers to crops, livestock, aquaculture and agroforestry.

FIGURE 1
Sources of investment in agriculture

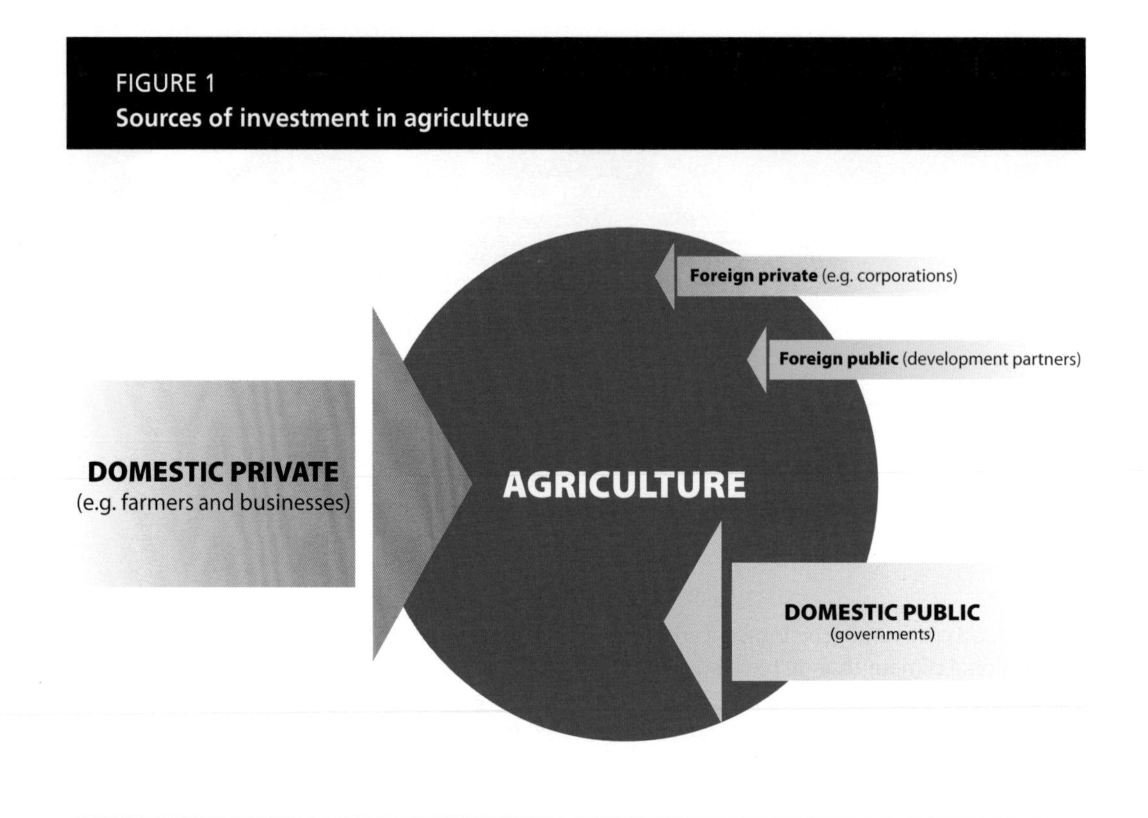

Source: FAO.

not substitutable for each other (Figure 1). The best available data, compiled and analysed for this report (Figure 5 in Chapter 2), only permit a rough comparison of the relative magnitudes of these investment flows, but the comparison highlights the central importance of farmers as the largest investors in agriculture. This has important implications for policy: while public investment remains essential, the focus of investment policy has to shift to facilitating more and better private investment.

Why invest in agriculture?

Farmers invest to feed their families, to increase and diversify their incomes and to build their wealth. For farmers, investing in agriculture means giving up something now (such as money, effort or time) in order to accumulate assets or capital that will allow them to increase their productivity and incomes in the future. Purchasing a plough, building an irrigation ditch, learning a new skill or nurturing trees and animals to reach a productive age are all forms of investment aimed at increasing the farmer's productivity or income. Farmers and other private investors will invest in agriculture only if the expected returns compensate for the perceived risk and exceed returns from alternative types of investment.

The rationale for public investment in agriculture by governments and development partners rests on three interrelated benefits for society that can come from enhancing agricultural productivity: (i) economic growth and poverty reduction, (ii) food and nutrition security, and (iii) environmental sustainability. For governments and donors, investing in agriculture means allocating scarce public resources to activities that raise productivity in the sector. Agricultural research and market infrastructure count among the most important types of public investment in agriculture.

History shows that even though farmers are the largest investors in agriculture, in the absence of good governance, appropriate incentives and essential public goods they do not invest enough.[2] Agricultural production is usually seasonal or cyclical in nature, and

[2] See Chapter 2 for clarification of basic concepts and Chapter 5 for a more detailed discussion.

is vulnerable to natural phenomena such as drought, pests and diseases. Producers are often geographically dispersed, and most agricultural products are bulky and perishable. All these factors make agricultural investment risky and highly dependent on the existence of good rural infrastructure, robust input supply and output processing industries, and transparent market institutions and price signals. Appropriate public investment can reduce the risk and increase the profitability of private investment and thus enhance incentives for farmers to invest.

An extensive body of evidence from many settings around the world shows that agricultural investment is one of the most important and effective strategies for **economic growth and poverty reduction** in rural areas, where the majority of the world's poorest people live. GDP growth in agriculture has been shown to be at least twice as effective in reducing poverty as growth originating in other sectors (World Bank, 2007a). Productivity growth in agriculture generates demand for other rural goods and services and creates employment and incomes for the people who provide them – often the landless rural poor.

These benefits ripple from the village to the broader economy in a process first documented decades ago (Hayami and Ruttan, 1970) and still valid in many rural areas today. Evidence presented in Chapter 5 shows that many of the most productive types of public investment for agriculture also have strong payoffs in terms of poverty reduction.

Agricultural investment is also key to eradicating hunger through the multiple dimensions of **food and nutrition security**. Investment by farmers and the public sector in agriculture and supportive sectors can increase the availability of food on the market and help keep consumer prices low, making food more accessible to rural and urban consumers (Alston *et al.*, 2000). Lower-priced staple foods enable consumers to improve their diets with a more diverse array of foods, such as vegetables, fruit, eggs and milk, which improves the utilization of nutrients in the diet (Bouis, Graham and Welch, 2000). Agricultural investments can also reduce the vulnerability of food supplies to shocks, promoting stability in consumption.

On-farm investment in agriculture appears to be closely linked to hunger reduction (Figure 2). Agricultural capital stock per

FIGURE 2
Average annual change in agricultural capital stock per worker and progress towards meeting the MDG hunger reduction target, 1990–92 to 2007

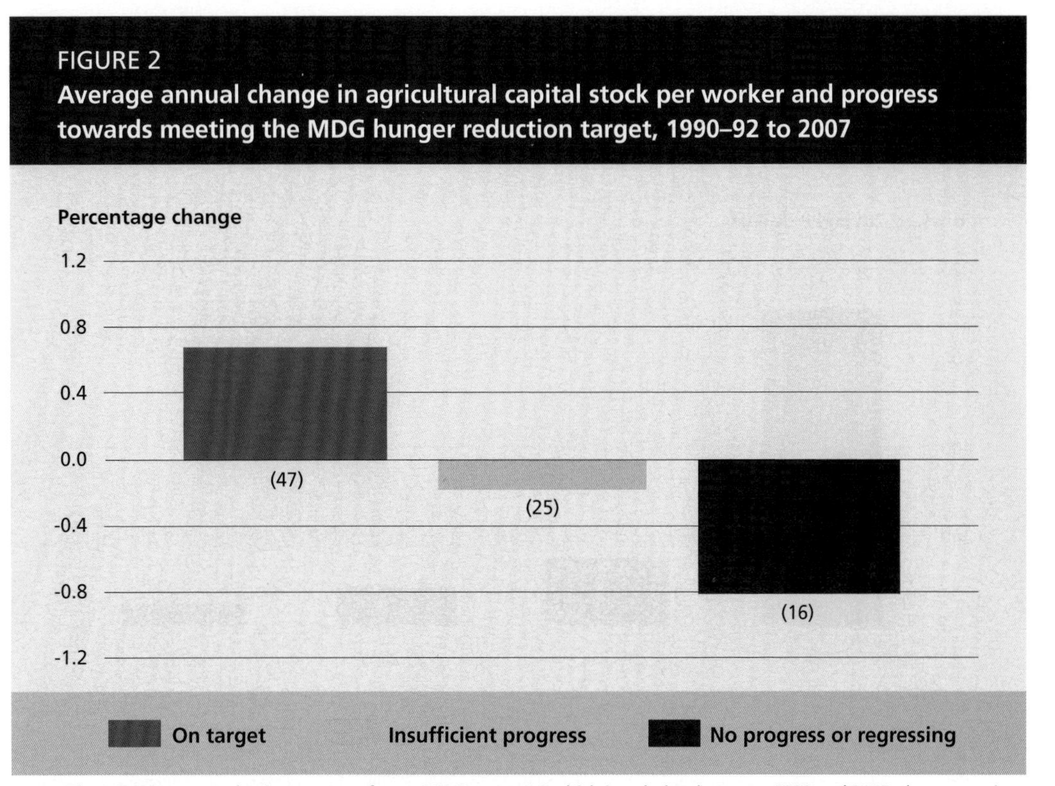

Notes: The MDG hunger reduction target refers to MDG target 1C which is to halve, between 1990 and 2015, the proportion of people who suffer from hunger. The number of countries in each category is shown in parentheses.
Source: Authors' calculations using FAO, 2012a and FAO, IFAD and WFP, 2012.

worker, a proxy for private domestic agricultural investment, has grown at an average rate of 0.7 percent per year since 1992 in the 47 countries that are on track to achieve the Millennium Development Goal's (MDG) hunger-reduction target, but it has declined slightly in the 25 countries where progress has been insufficient and strongly in the 16 countries where undernourishment rates have stagnated or regressed.

Private on-farm investment is clearly important for eradicating hunger, but public investment is also critical. Hunger is more prevalent in countries where public agricultural expenditure per worker is lower, suggesting that both public and private investment in agriculture are important in the fight against hunger (Figure 3). Of course, governments in low-income countries may spend less per agricultural worker precisely because they are poor, but evidence shows that many of them also spend proportionately less of their budgets on agriculture than is warranted by the prominence of agriculture in their economies (Chapter 2).

Productivity growth in agriculture is necessary – but not sufficient – to achieve **environmental sustainability**. World agriculture needs to feed a projected population of more than 9 billion people by 2050, some 2 billion more than today. Most of the population growth will occur in countries where hunger and natural resource degradation are already rife. Crop and livestock production systems must therefore become more intensive to meet growing demand but it will also be necessary to use fewer natural resources and improve the quality of these resources (FAO, 2011a). When agricultural ecosystems are more productive, natural ecosystems can be protected, and when farmers are rewarded for the value of the ecosystem services they provide, agriculture can become both more productive and more sustainable (FAO, 2007).

How to invest in agriculture for a better future?

Farmers in many low- and middle-income countries are not investing enough to meet their own goals of higher productivity and incomes, much less society's goals of food and nutrition security, poverty reduction

FIGURE 3
Government expenditures on agriculture per worker, by prevalence of undernourishment

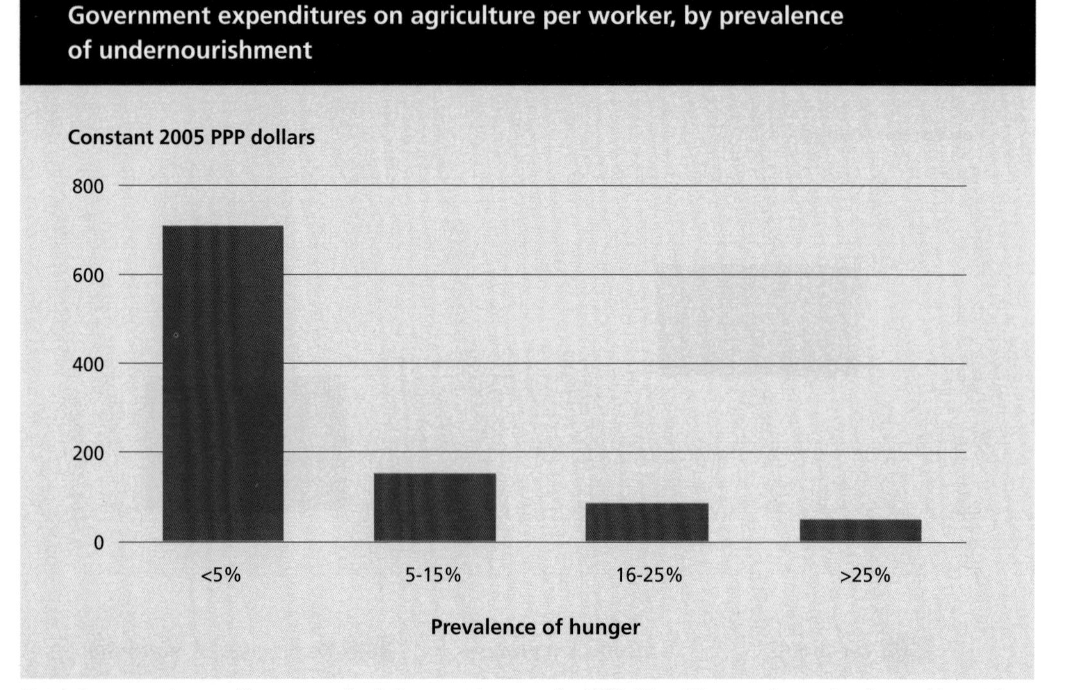

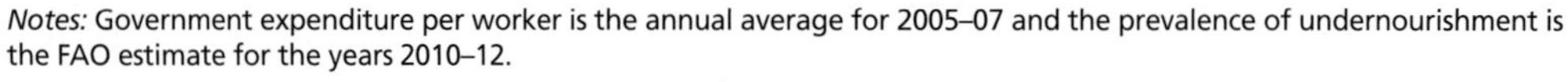

Notes: Government expenditure per worker is the annual average for 2005–07 and the prevalence of undernourishment is the FAO estimate for the years 2010–12.
Source: Authors' calculations using IFPRI, 2010 and FAO, IFAD and WFP, 2012.

and environmental sustainability. Addressing the incentives and constraints that influence farmers' investment decisions is imperative for stimulating on-farm investment.

Governments and their development partners have four basic responsibilities in this regard:

- create a conducive investment climate to catalyse socially responsible investment by farmers and other private investors;
- channel public expenditures towards the provision of essential, high-return public goods;
- overcome the constraints that smallholders face in saving and investing; and
- govern private investment, especially large-scale investment, to ensure social equity and environmental sustainability.

The relative importance of the four responsibilities and the priorities for public investment in and for agriculture will vary according to the level of economic development of the sector.

Governments have a major role in supporting a positive **investment climate** that is conducive to private investment in agriculture. The investment climate depends on the **enabling environment** – policies, institutions and infrastructure – for which governments are responsible and the **market incentives**, which are largely market-determined but are influenced by government policies in many domains. The investment climate influences the perceived profitability and risks associated with private investment, thus creating incentives or disincentives for farmers, rural enterprises and other private entities to invest in agriculture. The elements of an enabling environment and market incentives for investment in agriculture share many traits with a good general investment climate, although the relative importance of these elements may be different for agriculture.

The challenges faced by private and public investors in agriculture will vary according to context. Regional and country-level characteristics are influential, as are traits specific to the individual investor. However, all agricultural producers, regardless of their size or the country context, need the following basic features of an enabling environment: infrastructure and human resource development, trade and market institutions, macroeconomic stability and good governance. Agricultural investment is particularly dependent on such key enabling factors as predictability and transparency of policies, clear land tenure and property rights, transparent trade policy and physical rural infrastructure (including transportation, irrigation, communications, water and sanitation, and electric power). Other relevant enabling factors for agriculture include product norms and standards, research and development, and rural financial services (Chapter 3).

Many aspects of the enabling environment are **essential public goods**, which the private sector cannot be expected to provide. Governments have a responsibility to channel scarce public funds towards types of investment that have the highest payoff in terms of agricultural productivity, poverty reduction and environmental sustainability. Evidence presented in this report (Chapter 5) shows that public expenditures have higher social payoffs when they are concentrated on the provision of public goods such as agricultural research, rural infrastructure and education, rather than on subsidies for fertilizers, water and credit. Subsidies may be justifiable in some situations because they generate public good benefits; indeed, what constitutes a public good may differ according the level of development of the country. However, evidence is clear that some government expenditures have higher payoffs than others in terms of agricultural productivity and poverty reduction.

Governments also need to ensure that **environmental sustainability and social equity** considerations are effectively built in to private and public investment decisions in agriculture. This involves adopting laws and policies that support environmentally sustainable private investment and protect the rights of the most vulnerable. Policies in domains such as biofuel production, food self-sufficiency and international trade may have unintended adverse environmental consequences, which should be carefully evaluated. It also requires that public investment is directed towards enhancing production in ways that are environmentally sustainable and socially beneficial (Chapter 3).

In many countries, smallholders, many of whom are women, face particular constraints to saving and investing in their farms and may need special support in overcoming these. Linking smallholders to markets through appropriate institutions and infrastructure is part of an overall enabling environment and is a precondition for realizing the benefits accruing from better incentives. Overcoming credit constraints and risk aversion are other crucial challenges for smallholders. Helping build effective producer organizations can be a powerful way of linking smallholders to markets and overcoming some of the difficulties they face. In many contexts, social transfers, including subsidies, can also constitute an instrument that enables poor smallholders to invest and increase their assets (Chapter 4).

The increasing trend towards large-scale corporate investment in agriculture presents new opportunities and challenges for agriculture. Governments have a responsibility to govern such investment to ensure that it is conducive to food security and poverty alleviation in the countries and localities where it occurs. International organizations, civil society and corporate investors share the responsibility for governance of such investment. Adherence to the *Voluntary Guidelines on the Responsible Governance of Tenure of Land, Fisheries and Forests in the Context of National Food Security* (FAO, 2012b) and other rights-based principles are essential in this regard (Chapter 4).

The relevance and scale of the various policy challenges highlighted above will depend on individual country characteristics, level of development and priorities. Getting economic incentives right is critical for all countries – from low-income to high-income countries – as this has implications for geographic patterns of investment beyond the individual countries. Improving other elements of the investment climate is likely to be more challenging in many low- and middle-income countries. In the low-income countries and many lower-middle-income countries, with higher incidence of poverty and a large share of smallholders, addressing the constraints to smallholder investment and ensuring that large-scale investment is conducive to food security are crucial.

Investing in agriculture for a better future calls for a renewed partnership between governments, donors, civil society and the private sector – especially farmers – to ensure that significantly more investment is mobilized for agriculture and that it is channelled towards socially beneficial and environmentally sustainable outcomes. Building institutions and human capacity are central to this endeavour.

Structure of the report

Chapter 2 frames the debate by clarifying basic concepts related to agricultural investment and examining the empirical data on different types of investment. It reviews evidence on the importance of on-farm investment in agriculture as well as investment by governments, donors and private foreign investors. It highlights differences across regions and areas where investment may be lagging behind levels required to achieve sustainable productivity growth. Chapter 3 provides evidence on the crucial role of governments and donors in catalysing agricultural investment through the provision of an enabling environment and the transmission of price incentives. For example, macroeconomic and trade policies that tax or support the agriculture sector can influence incentives for investment in unintended ways. Furthermore, achieving sustainable intensification of agriculture requires the incorporation of environmental costs and benefits into the incentives available to agricultural producers. Chapter 4 gives special attention to the constraints to investment confronting smallholders and how governments and donors can help overcome them. The opportunities and challenges presented by recent trends towards large-scale corporate investment in developing country agriculture – by domestic and foreign investors – are also considered. Chapter 5 examines the returns on different types of public investment in different contexts and discusses how the reallocation of public expenditures towards essential public goods rather than subsidies can yield higher returns and socially more desirable outcomes. Chapter 6 draws conclusions and presents policy implications.

2. Agricultural investment: patterns and trends

Using data newly compiled and analysed for this report, this chapter reviews trends in private and public investment globally, regionally and by income group and assesses the extent to which progress is being made in agricultural capital formation within these areas.

Basic concepts: investment versus expenditures and public versus private goods

Broadly speaking, investment involves giving up something today in order to accumulate assets that generate increased income or other benefits in the future. Farmers invest in their farms by acquiring farm equipment and machinery, purchasing animals or raising them to productive age, planting permanent crops, improving their land, constructing farm buildings, etc. Governments may invest in, *inter alia*, building and maintaining rural roads and large-scale irrigation infrastructure, assets that generate returns in terms of increased productivity over a long period of time. Governments also invest in other, less tangible, assets such as the legal and market institutions that form part of the enabling environment for private investment. Determining whether an expenditure, public or private, constitutes an investment can thus be difficult both conceptually and empirically, and in some cases it is not clear-cut. Investment is generally defined as activities that result in the accumulation of capital (Box 1) that yields a stream of returns over time.

In agriculture, a distinction is usually made between investment and spending on inputs, based rather arbitrarily on the length of time required to generate a return. Thus, planting trees is typically considered an investment because it takes more than a year to generate a return, but applying fertilizer to a maize crop is not considered an investment because it generates a return during the current crop cycle. More important from a conceptual

BOX 1
What is capital?

Farmers and governments invest to build assets that promote agricultural productivity and growth. Capital is composed of both tangible and intangible assets and is often considered in terms of the following categories, all of which are important for agricultural productivity:

- Physical capital, such as animals, machinery, equipment, farm buildings, off-farm infrastructure;
- Human capital acquired through education, training and extension services;
- Intellectual capital acquired through research and development of agricultural technologies and management practices;
- Natural capital, such as land and other natural resources required for agricultural production;
- Social capital, such as the institutions and networks that build trust and reduce risk; and
- Financial capital, such as private savings.

Financial capital is primarily a means for acquiring other types of capital. However, many investments by farmers are not made primarily or exclusively through financial outlays but through time spent, for example in clearing or improving land or in constructing farm buildings or irrigation channels.

point of view, trees are a capital asset that yields a stream of returns over many years. Even in this seemingly simple case, the distinction may not be clear. If fertilizer use helps maintain and build soil fertility in the long run, it may also be considered an investment. Similarly, in public expenditures, a distinction is generally made between investment and current expenditures, but again this is not always clear-cut, not least because current expenditures are required to maintain the value of capital assets such as roads and other physical infrastructure.

Perspective also matters for what is perceived as investment. From a farmer's point of view, the purchase of land may represent an important investment in his or her productive capacity; from the perspective of society it simply involves a change in ownership of an asset rather than a net increase in capital stock, as occurs for instance when land improvements are undertaken.

Farmers and governments invest to build capital that allows the agriculture sector to become more productive in the future. Some of the most important types of capital for agriculture are not necessarily tangible. Governments invest extensively in agricultural research and development (R&D), which generates intellectual capital – a crucial input for raising the long-run productivity of agriculture. Both governments and individuals invest in education, which raises the productivity of the beneficiaries and generates long-term returns through human capacity development. Farmers spend time and resources developing producer associations, a form of social capital that can reduce risk and enhance productivity. All these activities are forms of investments because they build capital, even though the value of the capital may be difficult to measure.

Many of the investments made by governments are called "public goods" because they generate benefits for society that cannot be captured by a private investor. Once a public good has been created, people cannot be excluded from taking advantage of it, and use by one person does not diminish the ability of others to use it. In technical terms they are "non-exclusive" and "non-rival". Private investors have little or no incentive to provide public goods because they cannot charge enough to recover the cost of the investment. Examples of important public goods for agriculture include many types of R&D and rural roads and other infrastructure. Other types of public investment, such as building institutions and human capacity, provide less tangible but perhaps even more important public goods for agriculture. What constitutes a public good will depend to some extent on country characteristics and local context, and mixed public/private goods are common in agriculture.

Public investment helps create an appropriate enabling environment that influences farmers' incentives to invest. It also directly creates other forms of capital that support the development of a thriving agriculture sector. Some types of government investment are specific to agriculture and aimed specifically at enhancing primary production in the crop, livestock, aquaculture and forest sectors as well as in upstream and downstream activities. These can be referred to as investments *in* agriculture. Government investment in other sectors can also have a positive impact on agricultural production and productivity and on farm incomes. For example, investments in transport and communications infrastructure, energy, general education, health and nutrition, ecosystem services, market institutions and broader legal and social institutions all support agriculture and can be considered as investments *for* agriculture.

This report focuses on the accumulation of capital by farmers in agriculture and the investments made by governments to facilitate this accumulation. It does not cover the full range of investment in upstream and downstream private enterprises. Investment by input suppliers and agro-processors, for example, is crucial to supporting on-farm investment and agricultural development because it influences the opportunities and incentives perceived by farmers. Unfortunately, comprehensive data are not available for these sectors and they are outside the scope of the analysis, beyond noting their role in catalysing on-farm investment.

From concepts to measurement: making sense of the data

Moving from a conceptual understanding of agricultural investment to an empirical analysis poses a number of challenges because the available data provide only rough proxies for the components we want to measure. Despite some limitations, the data compiled and analysed for this report provide the most comprehensive and comparable estimates of investment in agriculture in low- and middle-income countries that have been prepared to date (Lowder, Carisma and Skoet, 2012).

Four key categories of investment and five internationally comparable data sets are analysed in this report (Figure 4). As noted in Chapter 1, the four categories of investment are domestic private, domestic public, foreign private and foreign public. Domestic private investment comes primarily from farmers, and the most comprehensive data available to measure this are estimates of on-farm agricultural capital stock calculated by FAO. Domestic public investment by governments is measured by two datasets: public expenditures on agricultural R&D from the Agricultural Science and Technology Indicators (ASTI) database (IFPRI, 2012a) and government expenditures in and for agriculture from the SPEED database (IFPRI, 2010 and IFPRI, 2012b), both maintained by the International Food Policy Research Institute (IFPRI). Both datasets measure aspects of public investment in agriculture. The best available measure of private foreign investment in agriculture and related sectors comes from data on foreign direct investment (FDI) compiled by the United Nations Conference on Trade and Development (UNCTAD). Foreign public investment is measured by data on official development assistance (ODA) to agriculture collected by the Organisation for Economic Co-operation and Development (OECD). None of these datasets captures the full range of asset accumulation in and for agriculture, but they are the most complete available.

The data clearly show that farmers are by far the largest investors in agriculture (Figure 5). On-farm investment is more than three times as large as all other sources of investment combined. Annual investment in on-farm agricultural capital stock exceeds government investment by more than 4 to 1 and other resource flows by a much larger margin. Agricultural capital stock measures only the most tangible forms of investment by farmers (i.e. land development, livestock, machinery and equipment, plantation crops [trees, vines and shrubs yielding repeated products] and structures for livestock). Because it excludes other forms of investment (e.g. education, training and participation in social networks), it probably represents a lower bound estimate of farmers' investment. Government investment is that portion of public expenditures that can be considered as investment (Box 5). In contrast, the R&D, ODA and FDI figures reported here do not distinguish between investment and current expenditures and thus represent an upper-bound estimate of these sources of investment.

Agricultural capital stock

Trends in total on-farm agricultural capital stock

The total accumulated investment by farmers worldwide, as measured by the value of agricultural capital stock, has increased about 20 percent since 1975 and now exceeds US$5 trillion (Annex table A2). At the global level, trends in total agricultural capital stock have been influenced by major political and economic events as well as international commodity prices (Figure 6). Sharply declining commodity prices throughout most of the 1980s and 1990s and unsupportive government policies provided fewer incentives for agricultural investment during this period.

The build-up of commodity stocks in the 1980s and early 1990s depressed investment in the high-income countries of Europe and North America. The collapse of the Union of Soviet Socialist Republics and economic reforms in the transition countries of Central and Eastern Europe led to sharp declines in agricultural capital stock in those countries during the 1990s. High rates of taxation of the agriculture sector further depressed investment in many low- and middle-income countries (see Chapter 3 for a more complete discussion). Progressive trade liberalization since the mid-1990s, following the completion of the Uruguay Round of multilateral trade negotiations, and higher commodity prices have improved the economic incentives to invest in agriculture through the mid-2000s. Continued high international commodity prices may have further stimulated investment in recent years, although comprehensive data to confirm this are not yet available.

INVESTMENT

FIGURE 4
Key international datasets on financial flows to agriculture

	DOMESTIC		
	PRIVATE	PUBLIC	
	On farm agricultural capital stock	**Government expenditures**	**Public spending on agricultural research and development**
Source	FAO	IFPRI-SPEED	IFPRI-ASTI
Sectors included	Crops and livestock	Crops and livestock	Crops and livestock, forestry, fisheries, natural resources, and on-farm food-processing
Definition	• Land development • Livestock • Machinery and equipment • Plantation crops (trees, vines and shrubs yielding repeated products) • Structures for livestock	• Administration supervision and regulation • Agrarian reform, agricultural land settlement, development and expansion • Flood control and irrigation • Farm price and income stabilization programmes • Extension, veterinary, pest control, crop inspection and crop grading services • Production and dissemination of general and technical information on agriculture • Compensation, grants, loans or subsidies to farmers	• Research on crops, livestock, forestry, fisheries, natural resources and socio-economic aspects of primary agricultural production • Research on on-farm postharvest activities and food-processing
Country coverage	204 countries and former sovereign states	Complete coverage for 51 countries, partial coverage for an additional 28 countries	140 countries in 2000, fewer in more recent years
Time span	1979–2007	1980–2007	1980– 2002 or 2009 (varies by country)
Unit of measure	Constant 2005 US$	Constant 2005 PPP dollars	Constant 2005 PPP dollars

FOREIGN

PRIVATE	PUBLIC
Foreign direct investment inflows	**Official development assistance**
UNCTAD	OECD-CRS
Crops and livestock, forestry, fisheries and hunting	Crops and livestock, forestry and fisheries
• Crops, market gardening and horticulture • Livestock • Mixed crops and livestock • Agricultural and animal husbandry services, excluding veterinary activities • Hunting, trapping and game propagation • Forestry and logging • Fishing, fish hatcheries and fish farms	• Agrarian reform, agricultural policy, administrative management, crop production, land and water resources, inputs, education, research, extension, training, plant and postharvest protection and pest control, financial services, farmers' organizations and cooperatives • Livestock production and veterinary services • Forestry policy and administrative management, development, production of fuelwood and charcoal, education and training, research and services • Fishing policy and administrative management, fisheries development, education and training, research and services
Varies by year (44 countries in most recent years)	153 countries
1990–2008	1973–2010
Current US$	Constant US$

AGRICULTURE

BOX 2
Better data on agricultural investment for policy analysis

Empirical analysis of investments in agriculture is rendered difficult by the very limited availability of data. This report provides the most comprehensive overview to date of trends in agricultural investment and of the magnitude of different sources of investment. All the datasets reviewed shed light on important dimensions of agricultural investment, but they are far from providing a complete picture.

Improved data would significantly enhance the analysis of agricultural investment. Improvements could cover different dimensions: comparability and consistency of data, country and year coverage, more up-to-date information and inclusion of areas not yet covered by data or estimates. Better coordination and collaboration among different institutions collecting data in similar or related areas could help. Specific areas for improvement include the following.

- **Agricultural capital stock.** Existing data have broad country coverage; however, the set of assets covered is significant but not complete and the methodology applied cannot account for improvements in quality of assets. Alternative estimates based on national accounts are currently only possible for a limited number of countries (Box 4).
- **Government expenditure.** Data compiled by IFPRI provide the most comprehensive information on government expenditures in low- and middle-income countries, but country coverage is not complete. There is also discrepancy between these data and data from other sources for specific countries. Harmonization and improvement of data on public expenditures could lead to better and more comprehensive data for analytical purposes. Also, a better breakdown of agricultural expenditures and more information on how much

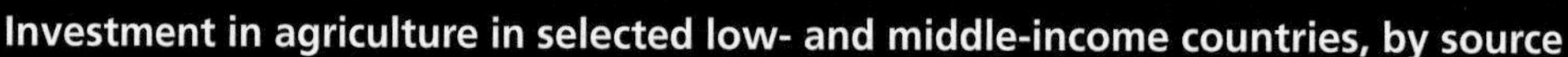

FIGURE 5
Investment in agriculture in selected low- and middle-income countries, by source

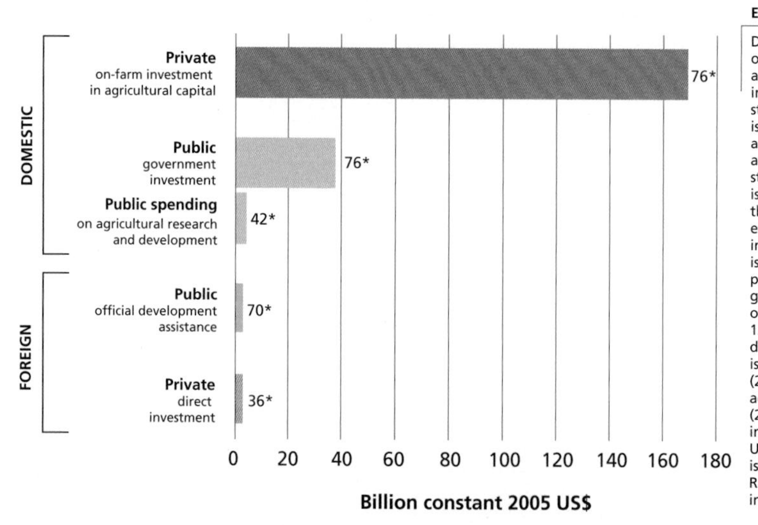

Explanatory note:

Data are averages for 2005–07 or for the most recent year available. Gross annual on-farm investment in agricultural capital stock (FAO, 2012a) is calculated using a 5 percent annual depreciation rate for the annual change in existing capital stock. Government investment is estimated using an assumption that 50 percent of government expenditures constitute investment. This assumption is based on a survey of agricultural public expenditure reviews, which give a mean of 42 percent for observations from a set of 12 countries (see Box 5). Official development assistance (ODA) is estimated using data from OECD (2012a); public spending on agricultural R&D is from IFPRI (2012a); and foreign direct investment (FDI) data are from UNCTAD (2011). No assumption is made regarding the share of R&D, ODA and FDI that constitute investment.

* Number of countries.

Source: Lowder, Carisma and Skoet, 2012.

they contribute to capital formation would improve the basis for analysis. Similarly, a breakdown of expenditure between rural and urban areas for types of non-agricultural investment that are strongly supportive of agriculture would also be important for analysis.

- **Research and development.** Data compiled by IFPRI's ASTI programme provide estimates of public expenditures – including government, higher-education, and non-profit – on agricultural R&D, but country coverage is limited and data are not updated with the necessary frequency to allow trends to be assessed over time. Funding for enhanced data collection would seem to be a priority. Also, private agricultural R&D appears to be a growing phenomenon in a number of low-and middle-income countries, but very limited information is available.
- **Foreign direct investment.** Data on FDI flows to agriculture are particularly weak. Available data are limited, inconsistent over time and far from comprehensive. One notable gap is the lack of coverage of investment by large institutional investors such as mutual funds, equity funds and pension funds, which appear to be growing.
- **Natural capital.** Natural resources are crucial for agricultural production and constitute some of the most important assets of developing countries. In spite of this, data assessing the value of natural resources for agricultural production are extremely limited.
- Finally, no internationally comparable data exist for **investment in value chains** beyond primary agriculture.

Agricultural capital stock per worker and labour productivity

More significant than the total level of agricultural capital stock is the amount per worker in agriculture,[3] because this is a major determinant of labour productivity and farm incomes (see Annex table A1 for data on the economically active population in agriculture). Figure 7 shows the correlation between agricultural capital stock per worker and labour productivity (measured by agricultural GDP per worker) for a large number of countries. Although the graphic cannot establish the direction of causality, the two are clearly highly correlated and rise markedly with overall per capita income levels. Broadly speaking, low-income countries have low levels of agricultural capital per worker and correspondingly low levels of agricultural output per worker. Low agricultural labour productivity may be considered a defining characteristic of low-income countries.

For agricultural labour productivity to grow, the amount of capital available for each worker (the capital–labour ratio) must grow. This requires agricultural capital stock to increase at a faster rate than the agricultural labour force. How quickly this occurs will affect the pace of farm income growth. In many instances, the gaps between high-income and low-income countries are widening as a result of low investment rates and/or growing labour forces in countries with low levels of agricultural capital per worker (Table 1). High rates of growth in the agricultural labour force have contributed both to declining capital per worker and declining farm size in the countries with the lowest levels of labour productivity (Box 3). Over the past decades, the capital–labour ratio has continued to increase rapidly in the high-income countries, primarily because

[3] Agricultural workers represent the economically active population in agriculture, including own-account farmers and formal or informal workers providing paid or unpaid labour.

FIGURE 6
Investment in agriculture and international commodity prices

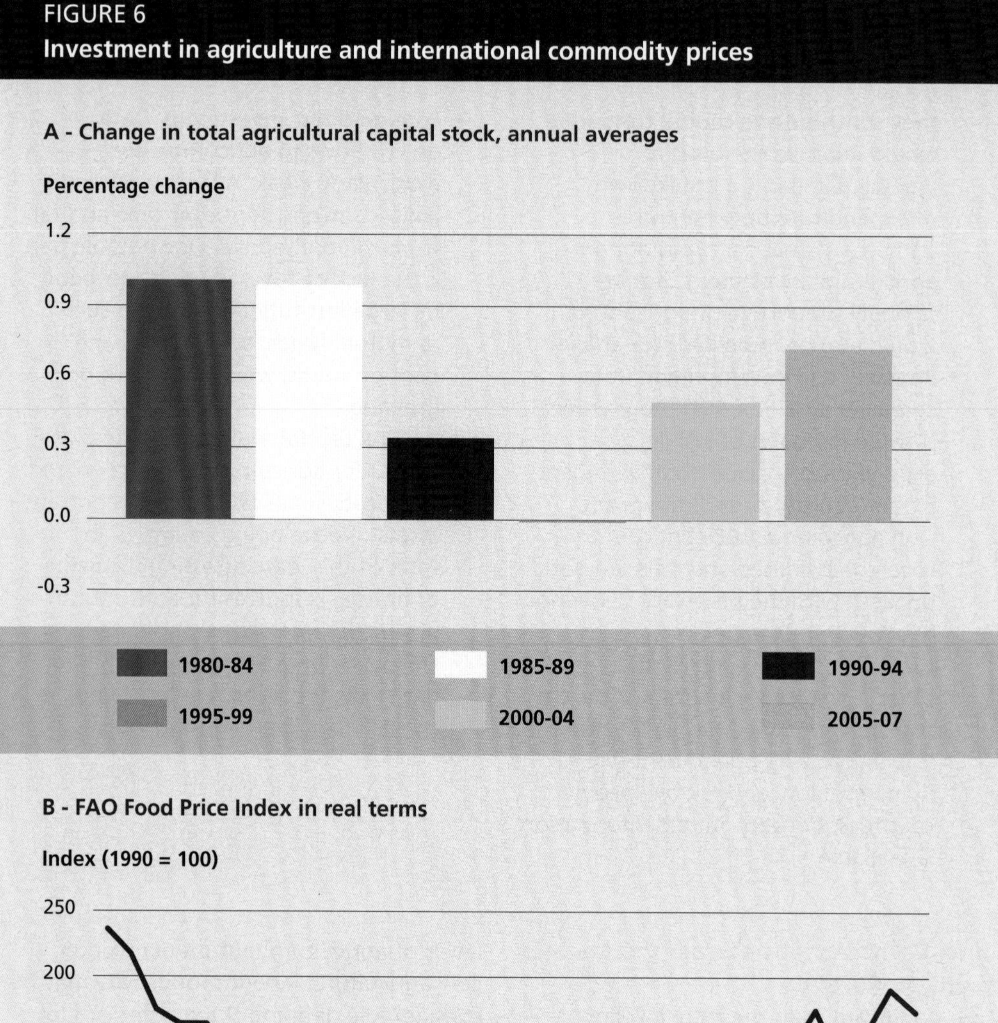

Note: The FAO Food Price Index is calculated using the international prices for cereals, oilseeds, meats and dairy products. FAO calculates it from 1990 to the present on a regular basis; in this figure it has been extended back to 1980 using proxy price information. The FPI for 2012 is calculated using data through May 2012. The index measures movements in international prices and not necessarily domestic prices. The United States GDP deflator is used to express the Food Price Index in real rather than nominal terms.
Sources: FAO Food Price Index: FAO, 2011b; change in total agricultural capital stock: authors' calculations using FAO, 2012a.

of falling numbers of workers in the sector, while it has declined in the low-income country group.

Regional trends in capital–labour ratios, are striking (Figure 8, page 19). Two regions in particular, with already low levels of capital per worker, saw stagnant or declining capital–labour ratios over three decades. In sub-Saharan Africa, where rapid growth in the agricultural labour force outpaced growth in total agricultural capital stock, the ratio fell at an average annual rate of 0.6 percent. In South Asia, the capital–labour ratio stagnated as total agricultural capital stock and the agricultural labour force grew at about the same rate.

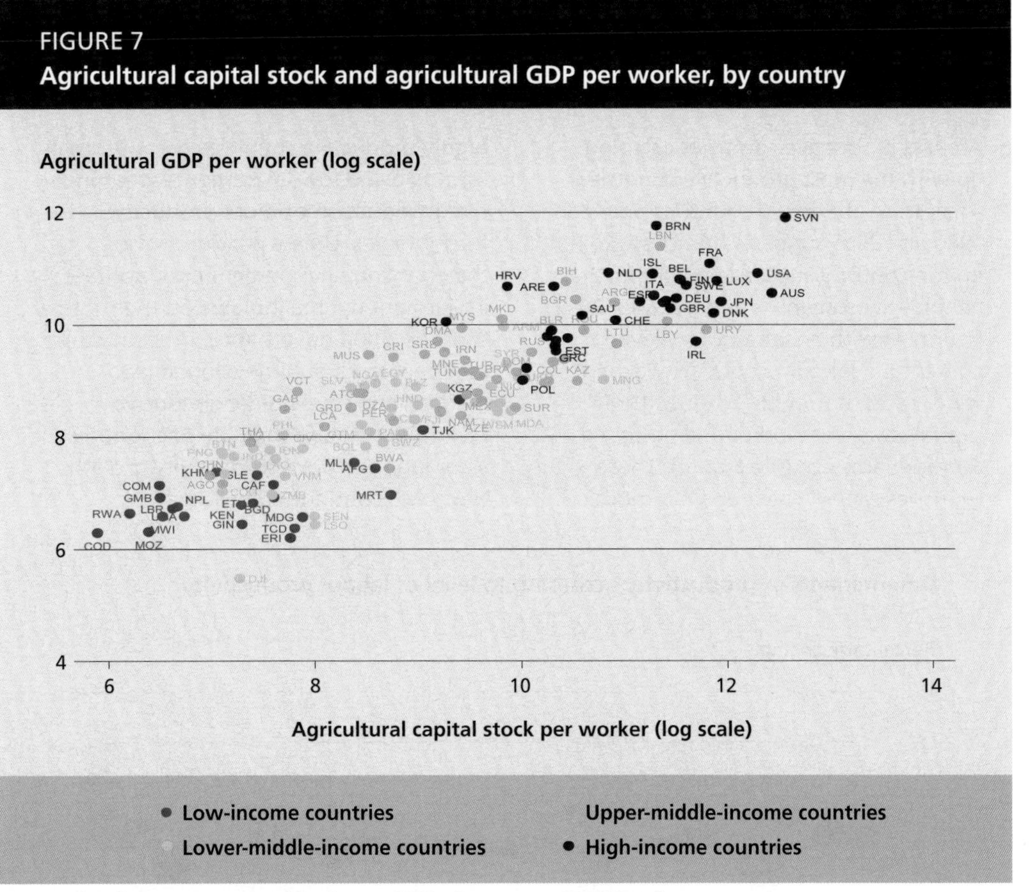

FIGURE 7
Agricultural capital stock and agricultural GDP per worker, by country

Notes: Both indicators are measured for the year 2007 using constant 2005 US dollars.

Sources: Authors' calculations using agricultural GDP data from the World Bank, 2012 and agricultural capital stock data from FAO, 2012a. See Annex table A2.

TABLE 1
Level and change in agricultural capital stock per worker, by region

INCOME GROUP/REGION	AVERAGE AGRICULTURAL CAPITAL STOCK PER WORKER, 2005–07	AVERAGE ANNUAL CHANGE (1980–2007) IN:		
		Agricultural capital stock	Number of agricultural workers	Agricultural capital stock per worker
	(Constant 2005 US$)		*(Percentage)*	
High-income countries	**89 800**	**0.2**	**-2.9**	**3.0**
Low- and middle-income countries	**2 600**	**0.9**	**1.2**	**-0.3**
East Asia and the Pacific	1 300	1.8	1.1	0.7
East Asia and the Pacific, excluding China	2 000	2.1	1.4	0.7
Europe and Central Asia	19 000	-1.0	-1.7	0.7
Latin America and the Caribbean	16 500	0.7	0.0	0.7
Middle East and North Africa	10 000	1.8	0.9	0.9
South Asia	1 700	1.4	1.4	0.0
South Asia, excluding India	3 000	1.4	1.6	-0.1
Sub-Saharan Africa	2 200	1.5	2.1	-0.6
WORLD	**4 000**	**0.6**	**1.1**	**-0.5**

Source: Authors' calculations using FAO, 2012a and World Bank, 2012. See Annex table A2.

BOX 3
The productivity gap

Are less productive countries catching up with the most productive countries? Analysis of about 100 countries between 1980 and 2005 suggests that they are not; on the contrary, most are falling further behind (Rapsomanikis and Vezzani, 2012). Countries with an initially low level of agricultural labour productivity exhibit lower rates of growth in agricultural capital stock per worker and declining average farm size (see Figure). These countries cannot catch up with more highly productive countries because small farm size and low investment rates hinder the introduction of more productive technologies. Unless policies provide the enabling environment and facilitate investment by smallholders on their farms, through good governance, infrastructure improvements, well-developed land markets and smallholder-conducive technology, the probability of countries escaping the "slow productivity growth trap" will continue to be low.

Determinants of productivity according to level of labour productivity

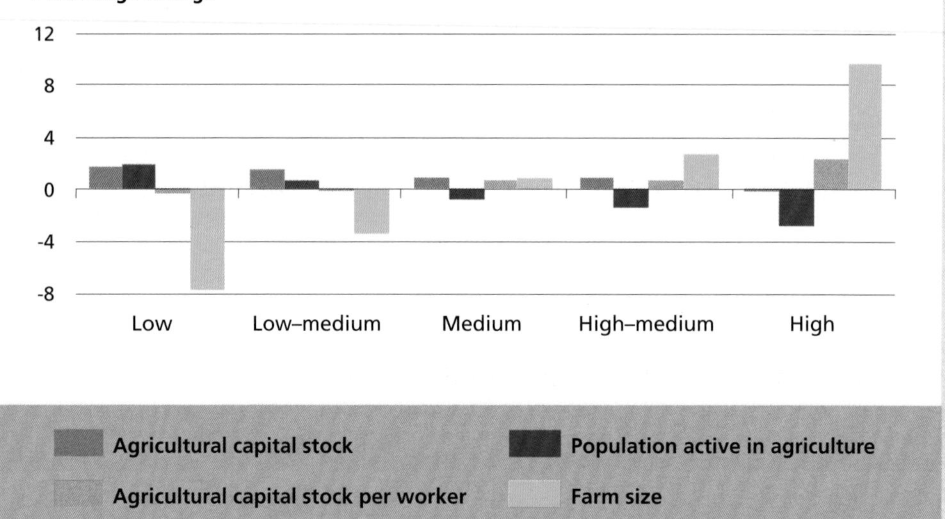

Note: Median annual growth rate, 1980–2005. Country groups are defined by quintiles in terms of labour productivity expressed as agricultural GDP per economically active worker in the sector. Each group represents 20 percent of the sample.
Source: Authors' calculations using FAO, 2012a.

The composition of agricultural capital stock

The composition of agricultural capital stock has implications for agricultural labour productivity and environmental sustainability. Natural resources (a major component of natural capital) constitute some of the most important assets of developing countries and they form the biophysical foundation for agriculture. The World Bank (2006a) estimated that natural capital represented about 26 percent of the total wealth of low-income countries (excluding oil states) in 2000 – a greater share than produced capital (infrastructure, buildings, machinery and equipment) at 16 percent. Cropland constituted by far the largest share (59 percent) of natural capital, with subsoil assets (17 percent) and pastureland (10 percent) accounting for the next largest shares. The relative share of natural capital is lower for countries with higher income levels, amounting to 13 percent in the middle-income countries and 2 percent in the high-income countries.

FIGURE 8
Average annual change in agricultural capital stock per worker in low- and middle-income countries, 1980–2007

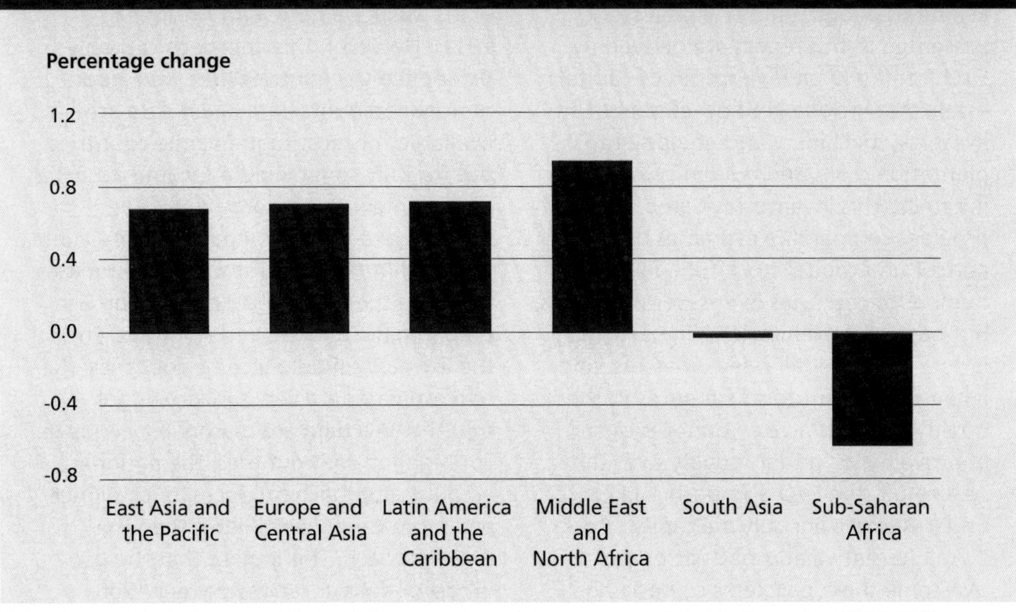

Notes: For countries in Europe and Central Asia, average annual changes are calculated for the period 1992 to 2007.
Source: Authors' calculations using FAO, 2012a and World Bank, 2012.

Despite the importance of natural capital, data on many aspects of natural capital – such as those relating to the quality of soils, water and genetic resources – are limited. Available measures of agricultural capital stock (such as FAO's) thus rely on measures such as machinery, livestock, structures and land development.

As agriculture becomes technologically more advanced, the composition of agricultural capital changes. There are major differences in the composition of agricultural capital stock in the high-income countries and in the low- and middle-income countries, especially concerning the share of machinery and equipment (Figure 9, page 21). Machinery and equipment account for more than 40 percent of total agricultural capital stock in the high-income countries, in stark contrast with less than 3 percent in the low-income countries. For the low- and middle-income countries, the dominant forms of on-farm capital are those embodied in livestock and land improvements.

Sustained productivity gains over time depend on changes in capital, including those aspects of natural capital for which data are scarce. Sustainable production systems are also knowledge-intensive, so the transition to sustainable, climate-smart agriculture will imply a greater reliance on types of capital that embody intellectual and human capital in order to economize on increasingly scarce natural resources. Available measures of agricultural capital stock only partially capture knowledge-related capital (machinery and equipment are one proxy, but very crude and incomplete). A key conclusion is that investment is needed in precisely the kinds of assets that are becoming most relevant to decision-making about sustainable productivity growth, namely the quality of natural and human capital – as well as in the activities, such as agricultural R&D, that can help improve them.

Implications of trends in agricultural capital stock

The trends in agricultural capital stock, agricultural capital stock per worker and the composition of agricultural capital stock all suggest that investment is seriously lagging in the low- and lower-middle-income countries, and particularly in sub-Saharan Africa and South Asia. The close correlation between capital–labour ratios and agricultural labour productivity suggest that significant increases in on-farm investment will be required in these regions in order

BOX 4
Alternative estimates of agricultural capital stock

Estimates of agricultural capital stock presented in this report are derived by FAO from data on inventories of capital assets that include land development, livestock, machinery and equipment, plantation crops and buildings for livestock. This inventories-based approach provides comparable estimates of agricultural capital stock for a large number of countries over several decades, but has various limitations; in particular, it does not cover all relevant assets, and it cannot account for differences in the quality of assets across countries or for improvements in their quality over time. As a result, the FAO approach is likely to underestimate agricultural capital stock.

An alternative approach attempts to overcome these problems by deriving estimates of agricultural capital stock from investment data reported in national accounts (Crego *et al.*, 1997; Larson *et al.*, 2000; Daidone and Anríquez, 2011). However, this approach can only be applied to countries that have good national accounts data. Such data are available for most high-income countries but for only some middle-income countries and very few low-income countries.

The figure below compares the FAO data on agricultural capital stock with estimates based on the national accounts approach prepared by Daidone and Anríquez. For the low- and middle-income countries, the two estimates are very similar, suggesting that the FAO data are reasonably accurate. For high-income countries, the national accounts approach produces much higher and more variable estimates than the FAO approach. This implies that the gap in capital–labour ratios between high-income countries and low- and middle-income countries may be even wider than indicated by the FAO data.

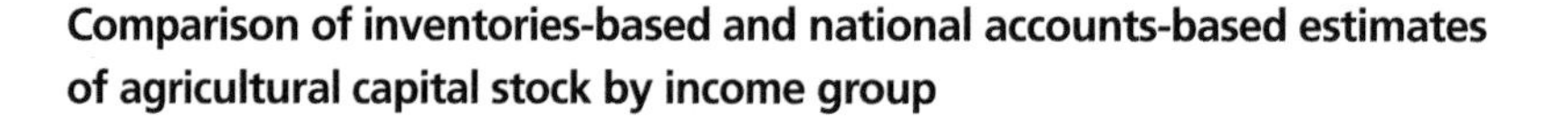

Comparison of inventories-based and national accounts-based estimates of agricultural capital stock by income group

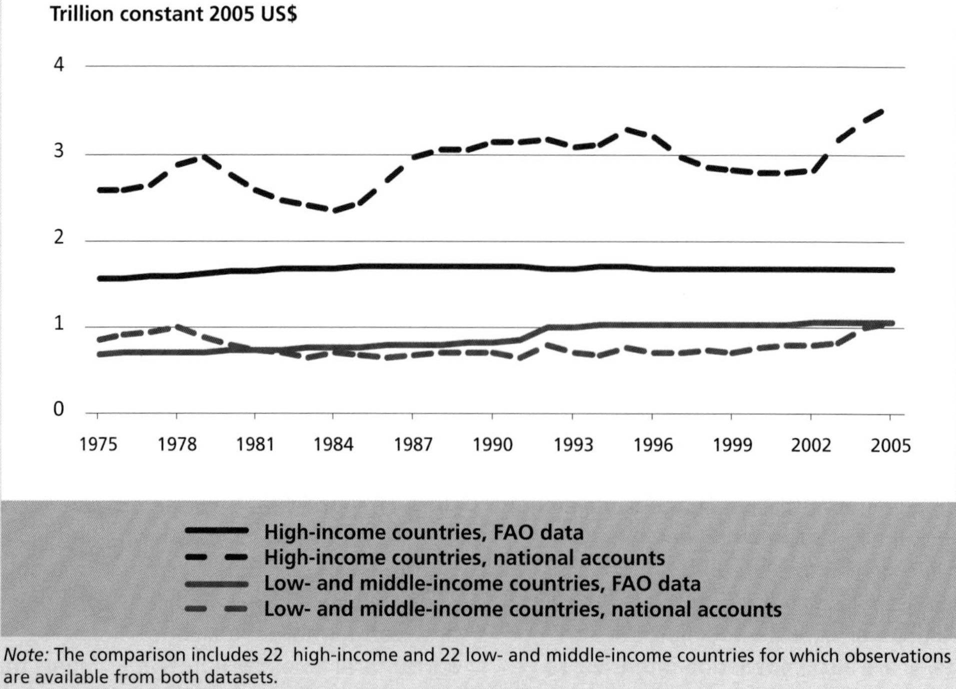

Note: The comparison includes 22 high-income and 22 low- and middle-income countries for which observations are available from both datasets.
Source: Authors' calculations using FAO, 2012a and Daidone and Anríquez, 2011.

FIGURE 9
Composition of agricultural capital stock by income group, 2005–07

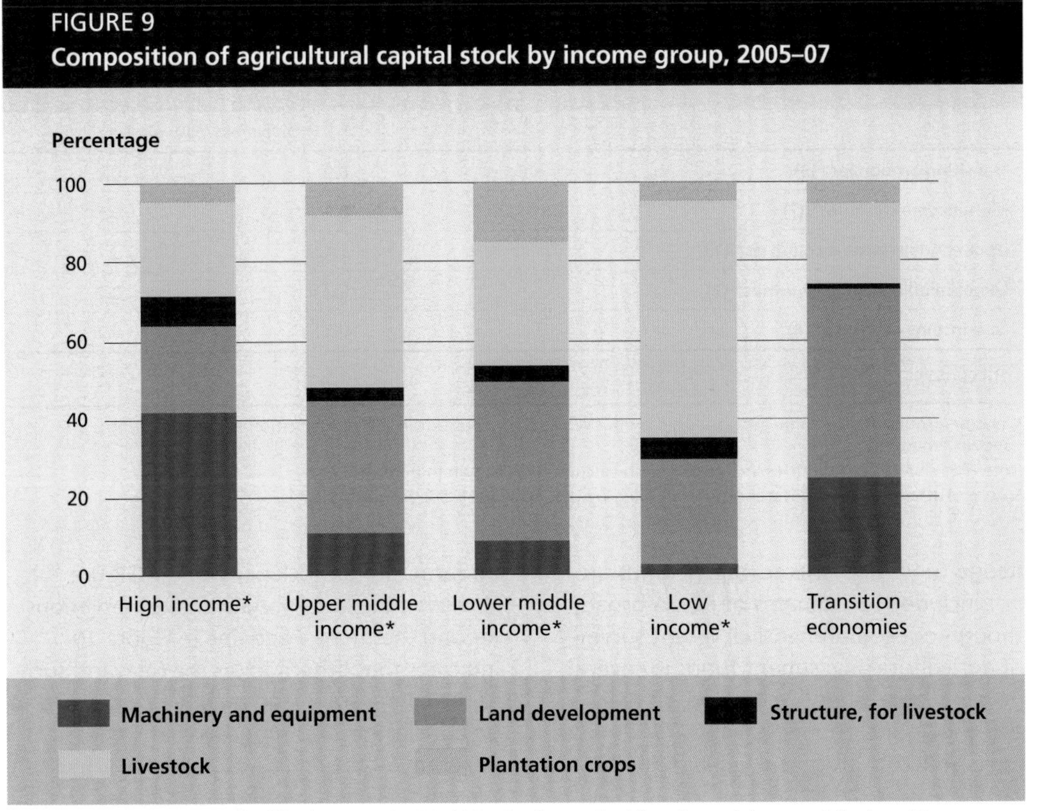

Note: *The income group classification uses the World Bank atlas method for all countries except the transition economies, which are presented as a separate group.
Source: Authors' calculations using FAO, 2012a.

to make progress against poverty, hunger and resource degradation. Broader changes in the agricultural economy, including a transition of labour out of the sector as has occurred in other regions as a result of economic growth, will also be necessary.

Foreign direct investment in agriculture

Much recent attention has been given to FDI, which appears to be a growing source of investment in agriculture in low- and middle-income countries. Data limitations make it difficult to draw solid conclusions about the magnitude of such investment globally, or the long-term trends, but the best available data show that agricultural FDI remains very small compared with domestic agricultural investment (See Annex table A3 for data by country). In addition, it is unclear how much it contributes to capital formation as opposed to a mere transfer of ownership.

For 2007 and 2008, comparable data on total FDI to all sectors are only available for 27 countries. For these countries, average annual inward FDI flows in the two years were estimated at US$922.4 billion (UNCTAD, 2011). Of this total, FDI to agriculture (including hunting, forestry and fisheries) represented only 0.4 percent. A larger share, 5.6 percent, went to the food, beverages and tobacco sectors, primarily in high-income countries.

Trends over time in FDI are difficult to monitor because the number of countries for which data are available varies from year to year. Looking at agriculture alone, recent comparable data are available for 44 countries; FDI to these countries more than doubled between 2005–06 and 2007–08 (Table 2). However, the majority of these flows went to upper-middle and high-income countries (Lowder and Carisma, 2011).

These figures underestimate actual flows of foreign investment in agriculture, because data are missing for so many countries and only direct investment by private companies is included. Investments made by large institutional investors, such as mutual funds, banks, pension funds,

TABLE 2
Average annual foreign direct investment in agriculture, by income group

INCOME GROUP	2005–06	2007–08
	(Current US$, billions)	
Transition economies (13)	0.3	0.8
High-income countries* (7)	0.1	0.5
Upper-middle-income countries* (13)	1.4	3.7
Lower-middle-income countries* (7)	0.2	0.3
Low-income countries* (4)	0.1	0.2
Total (44)	**2.1**	**5.4**

* Income groups are the same as those used by the World Bank, but not including transition economies, which are shown separately.
Note: The number of countries included in each calculation is shown in parentheses.
Source: Authors' calculations using data supplied by UNCTAD, 2011. See Annex table A3.

hedge funds and private equity funds are not included in estimates of FDI. A broad, though not comprehensive, recent survey of agricultural investment funds in several developing regions (excluding East Asia and the Pacific) found that such funds have increased in number and value (Miller *et al.*, 2010).

However, given the relatively small size of FDI flows to primary agriculture reported in the international dataset, especially in low-income countries, it is unlikely that FDI can contribute significantly to raising capital stock in agriculture. Nevertheless, it can still have significant impacts at the local level. FDI in agriculture may offer opportunities for developing countries in terms of employment and technology transfer, but potentially negative social and environmental impacts of such investments (especially those that involve direct control of agricultural land) remain a reason for concern. The issue of foreign investment and land acquisition in developing countries will be examined more closely in Chapter 4.

Government expenditures on agriculture

After farmers' investment in on-farm capital stock, the second-largest source of investment in agriculture is government expenditures. Public expenditures constitute an essential component of creating an enabling environment for farm investment and are positively correlated with the formation of on-farm capital stock per worker (Figure 10). However, the large variation of observations around the fitted trend line in Figure 10 indicates that other factors are relevant, such as the composition and quality of expenditure on agriculture. This suggests that some government expenditures are more effective than others in promoting agricultural investment and growth.

Government expenditures have been growing in real terms over the last three decades in the 51 low- and middle-income countries covered by a database released by IFPRI (2010), but trends differ by region and income group (Figure 11; see also Annex table A4 for information by country). Agricultural expenditures grew more slowly than other expenditure categories, and the share of agriculture in overall government expenditures has consequently declined. The long-term decline in the share is common to all regions (Figure 12). Only South Asia seems to have seen a renewed increase in the share of agricultural expenditures in the most recent years. Not all government expenditure on agriculture constitutes investment and assessing how much of it contributes to capital formation is not straightforward (Box 5).

More important than overall levels of agricultural expenditure or their share in total government expenditure are measures that assess these trends relative to the role of agriculture in the economy. One such measure is government expenditures on agriculture per worker in the sector (Table 3; see Annex table A5 for data by country).

FIGURE 10
Government expenditure on agriculture and percentage change in agricultural capital stock per worker in selected low- and middle-income countries

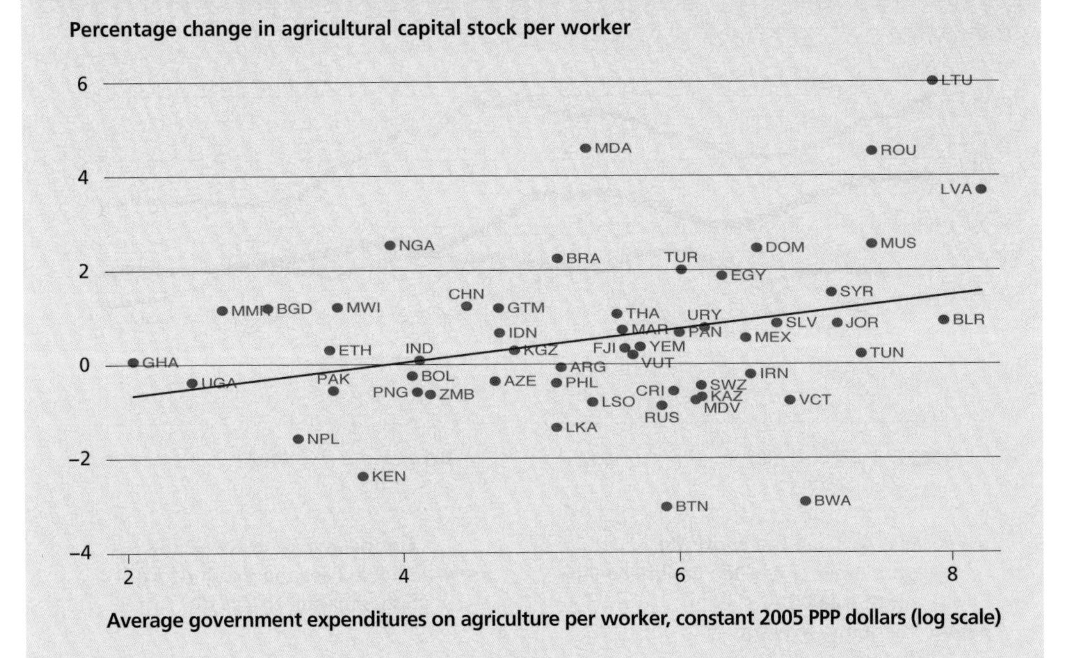

Note: Change in agricultural capital stock and government expenditures are annual averages from 1990 to 2007 for all countries except those located in Europe and Central Asia, for which averages are from 1995 to 2007.
Source: Authors' calculations using IFPRI, 2012b and FAO, 2012a.

FIGURE 11
Government expenditure on agriculture, by region

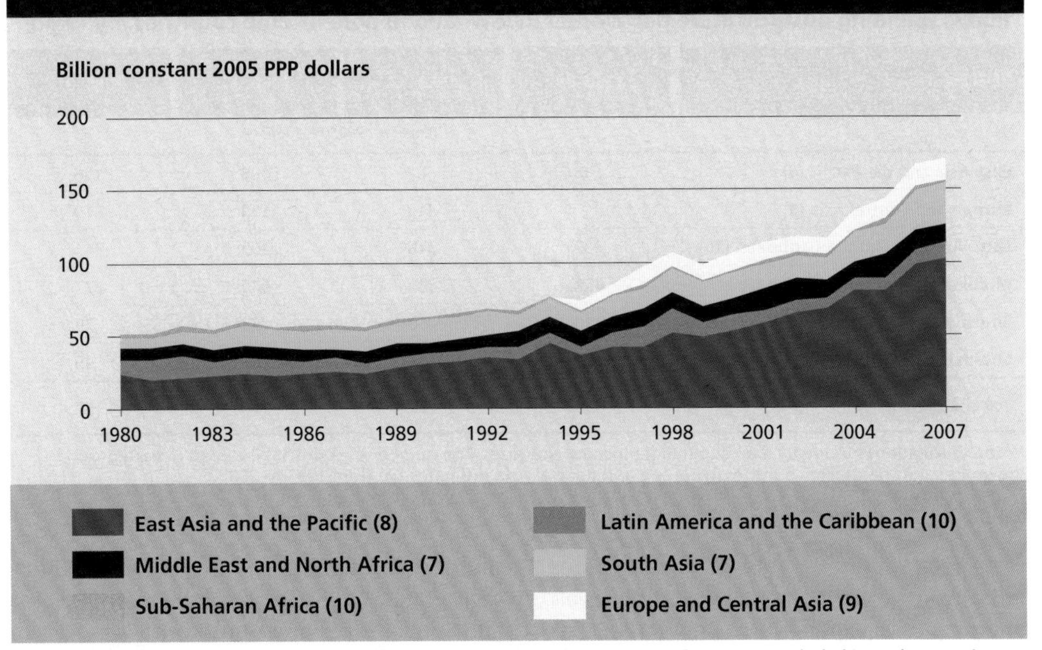

Note: Calculations include 51 low- and middle-income countries. The number of countries included in each group is shown in parentheses. For countries in Europe and Central Asia estimates are from 1995 to 2007.
Source: Authors' calculations using IFPRI, 2010. See Annex table A4.

FIGURE 12
Agricultural share of public expenditure, by region, three-year moving averages

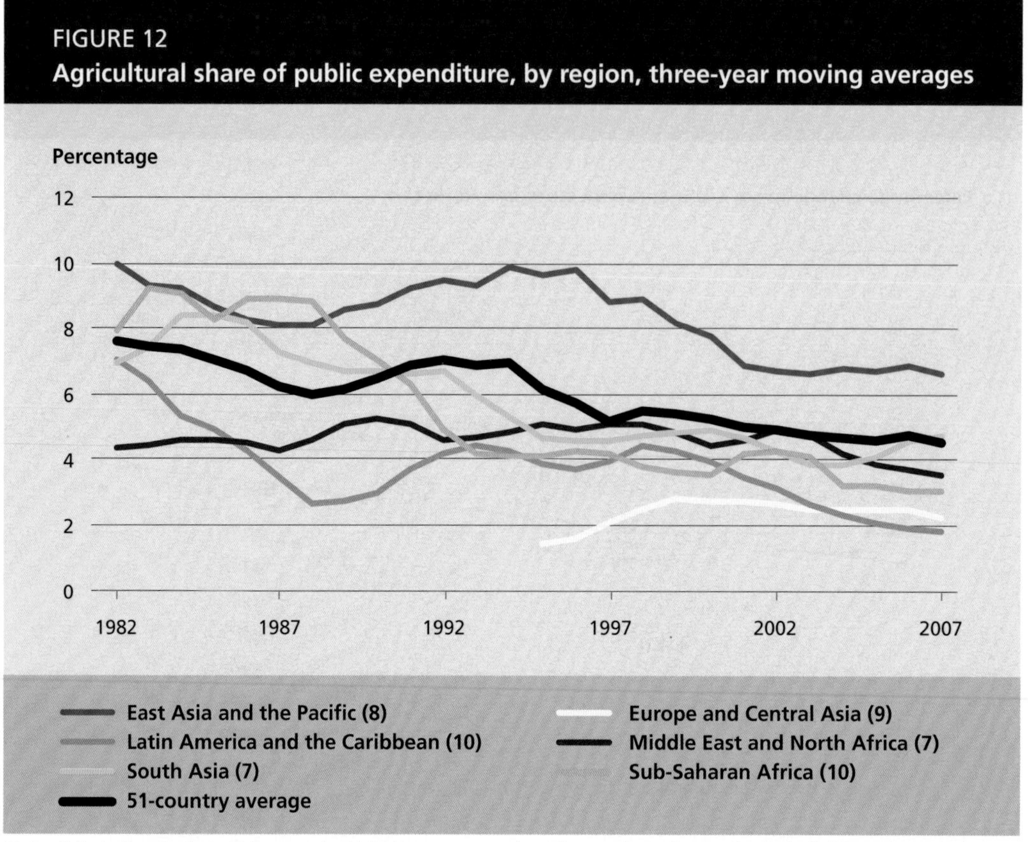

Note: Calculations include 51 low- and middle-income countries. The number of countries included in each group is shown in parentheses. For countries in Europe and Central Asia estimates are from 1995 to 2007. Ethiopia has been excluded from the calculation of the regional average for sub-Saharan Africa for this and other graphics and tables on government expenditure. According to the SPEED database, the share of agriculture in public expenditures in Ethiopia increased from 4–7 percent in 2001–04 to 14–17 percent in 2005–07.
Source: Authors' calculations using IFPRI, 2010. See Annex table A4.

TABLE 3
Public spending on agriculture per worker in low- and middle-income countries, by region

REGION	1980–89	1990–99	2000–04	2005–07
		(Constant 2005 PPP dollars)		
East Asia and the Pacific (8)	48	69	108	156
Europe and Central Asia (9)		413	559	719
Latin America and the Caribbean (10)	337	316	309	341
Middle East and North Africa (7)	458	534	640	677
South Asia (7)	46	50	53	79
Sub-Saharan Africa (10)	152	50	51	45
Total (51 countries)	**68**	**82**	**114**	**152**

Notes: Calculations include 51 low- and middle-income countries. The number of countries included in each group is shown in parentheses. For countries in Europe and Central Asia estimates are from 1995 to 2007.
Source: Authors' calculations using IFPRI, 2010 and FAO, 2012a. See Annex table A5.

From the 1980s to the late 2000s, all regions but one increased or maintained their levels of agricultural expenditures per worker. The conspicuous exception is sub-Saharan Africa, where spending per worker declined by more than two-thirds between the 1980s and the early 2000s. Countries in sub-Saharan Africa and South Asia spend significantly less per agricultural worker than those in any other region.

BOX 5

How much of public expenditure on agriculture is investment? Evidence from public expenditure reviews

It is not always easy to determine which government expenditures should be considered investment and which should not. Public expenditure reviews (PERs) are an important tool for assessing and analysing public expenditures and can provide a useful benchmark against which to evaluate the effectiveness of government expenditures. The content and format of such reviews vary, due to differences in purpose, approach and sectoral coverage, thus they may not allow the kind of cross-country comparability that would be needed in a international score card system. Some PERs for the agriculture sector available in the public domain provide information on the breakdown of agricultural expenditures, including by capital and current expenditures (see Table).[1] The share of capital expenditures in total expenditures is highly variable, ranging from as little as 9 percent in the United Republic of Tanzania to 84 percent in Lao People's Democratic Republic and Mozambique. In some cases, a clear difference is also recorded between budgeted and actual expenditures.

[1] The terms " current (or recurrent) expenditures" and "capital expenditures" are frequently found in the economics literature analysing public expenditures, including public expenditure reviews, but are not used in the formal manuals and guides on government statistics. The International Monetary Fund's *Government Finance Statistics Manual* (IMF, 2001) distinguishes between expenses and expenditures on (non-financial) assets and public capital formation. The two sets of concepts are close, but not identical.

Share of capital expenditures in overall agricultural expenditures from selected public expenditure reviews

COUNTRY	CAPITAL SHARE OF AGRICULTURAL EXPENDITURES *(Percentage)*	NOTES	PERIOD
Ghana (1)	17	Development, total (a)	2005
	24	MoFA, actual	
	46	MoFA, budgeted	
Honduras (2)	66		2006
Kenya (3)	30		2004/05
Lao People's Democratic Republic (4)	84		2004/05
Mozambique (5)	84	Total (b)	2007
	9	MINAG	
Nigeria (6)	58	Budgeted	2001-05
	44	Actual	
Nepal (7)	46	(c)	1999-2003
Philippines (8)	26	(d)	2005
Uganda (9)	24		2005/06–2008/09
United Republic of Tanzania (10)	9		2011
Viet Nam (11)	77		2002
Zambia (11)	24		2000

Notes: (a) Development as opposed to recurrent expenditures. Covers all government expenditure, as opposed to only those made by MoFA (Ministry of Food and Agriculture), the latter accounts for about 25 percent of total government expenditure in this sector. (b) 84 percent refers to total government expenditure; 9 percent is for MINAG (Ministry of Agriculture [Ministério da Agricultural]) only. (c) Includes irrigation and agriculture expenditures. (d) Consolidated Department of Agriculture expenditure figures.
Sources: (1) Kolavalli *et al.*, 2010; (2) Anson and Zegarra, 2008; (3) Akroyd and Smith, 2007; (4) Cammack, Fowler and Phomdouangsy, 2008; (5) World Bank, 2011a; (6) World Bank, 2008; (7) Dillon, Sharma and Zhang, 2008; (8) World Bank, 2007b; (9) World Bank, 2010a; (10) World Bank, 2011a; (11) Akroyd and Smith, 2007.

The Agricultural Orientation Index (AOI) provides a way to assess whether government expenditures on agriculture reflect the economic importance of the sector (Table 4, page 28; see Annex table A5 for data by country). This index is calculated as the share of agriculture in total government expenditure divided by the share of agriculture in total GDP. It is an indicator of the degree to which the share of agriculture in public expenditure is commensurate with the weight of the sector in GDP.[4] Time trends in the index vary across regions, but the most striking is that of sub-Saharan Africa, where the AOI is well below half the level it was in the 1980s.

Composition of public expenditures

As seen above, the decline in the share of agriculture in public expenditure is not generally the result of declining levels of expenditure on agriculture, but of larger increases in other areas that have been given higher priority over time. For a complete picture of the dynamics of public expenditures on agriculture, they must be seen in the context of the dynamics of overall government expenditure patterns (Table 5, page 28).

On average, governments in all regions currently spend more on defence than on agriculture. The share of education in public expenditure has also increased significantly since 1980 in all regions except the Middle East and North Africa, while all regions have seen an increase in the share spent either on health or social protection, if not both. All of these are expenditure categories with a significant potential development impact, and in many cases they are also likely to have a positive impact on agricultural and rural development. They may include significant levels of expenditures *for* agriculture. However, at the same time, the share of another expenditure category with a possible positive impact on agriculture – transport and communication – has declined over time in most regions.

Given fiscal constraints, increased public expenditures on agriculture would have to

[4] The AOI is useful for comparisons across countries and over time, but it is not prescriptive. Many essential government expenditures – such as education, health, infrastructure and social transfers – do not reflect the economic contribution of the relevant sector.

BOX 6
The 2003 Maputo declaration and the share of agriculture in government spending in African countries

At the Assembly of the African Union in July 2003 in Maputo, African Heads of State and Government endorsed the "Maputo Declaration on Agriculture and Food Security in Africa", which established the Comprehensive Africa Agriculture Development Programme (CAADP, see Box 23 on page 87). Two significant targets were to increase agricultural productivity by 6 percent annually through 2015 and to allocate at least 10 percent of national budgetary resources to agriculture and rural development within five years.

Notwithstanding whether 10 percent is necessarily the appropriate budgetary allocation to agriculture, such a target can provide a useful benchmark against which to evaluate a country's commitment to agriculture. The Regional Strategic Analysis and Knowledge Support System (ReSAKSS) – an Africa-wide network – was established to provide analytical tools to support policy-making and to evaluate progress towards the CAADP goals. The system compiles data on the share of government spending going to agriculture in African countries. As shown in the Figure, only seven countries covered by the data had attained the 10 percent target in the most recent year for which information is available.[1]

[1] There are discrepancies between the data from ReSAKSS and the SPEED database arising from differences in definitions, coverage and data sources. The variations from year to year can be significant, even for countries that have reached the target or progressed.

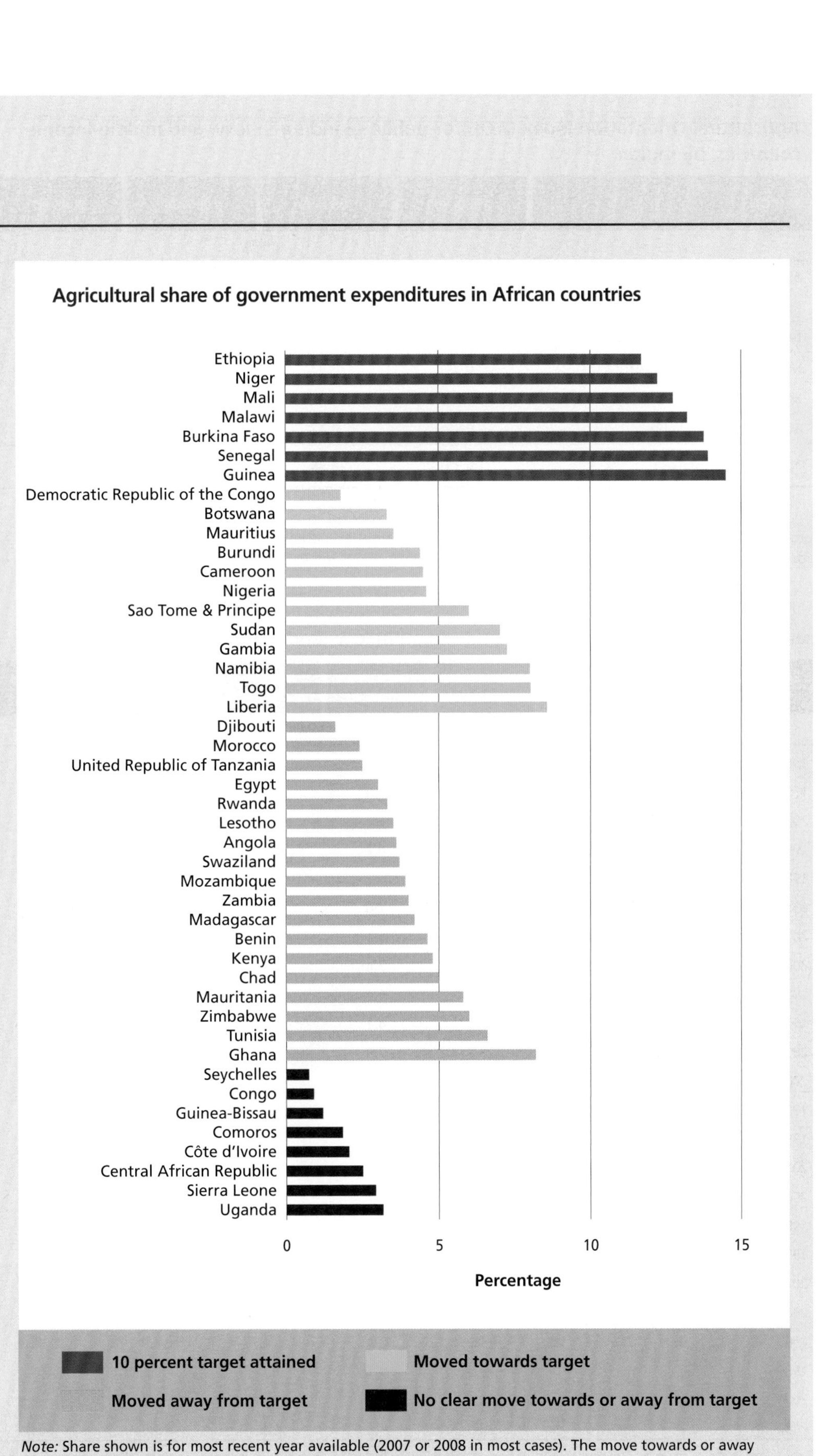

Note: Share shown is for most recent year available (2007 or 2008 in most cases). The move towards or away from the target is based on changes in the last three available years.
Source: Authors' calculations using ReSAKSS, 2011.

TABLE 4
Agricultural Orientation Index (AOI) for public spending in low- and middle-income countries, by region

REGION	1980–89	1990–99	2000–04	2005–07
			(Ratio)	
East Asia and the Pacific (7)	0.31	0.48	0.49	0.59
Europe and Central Asia (9)		0.29	0.35	0.36
Latin America and the Caribbean (6)	0.96	0.86	0.56	0.38
Middle East and North Africa (5)	0.34	0.37	0.37	0.30
South Asia (5)	0.24	0.21	0.21	0.27
Sub-Saharan Africa (9)	0.30	0.17	0.14	0.12
Total (41 countries)	**0.35**	**0.38**	**0.38**	**0.41**

Notes: The AOI for public spending equals the agricultural share of government spending divided by the agricultural share of GDP. Calculations include 41 low- and middle-income countries. The number of countries included in each group is shown in parentheses. For countries in Europe and Central Asia estimates are from 1995 to 2007.
Source: Authors' calculations using IFPRI, 2010 and World Bank, 2012. See Annex table A5.

TABLE 5
Composition of government expenditures, by sector and region in low- and middle-income countries

REGION	YEAR	AGRICULTURE	DEFENCE	EDUCATION	HEALTH	SOCIAL PROTECTION	TRANSPORT AND COMMUNICATION	OTHERS
					(Percentage share of total)			
East Asia and the Pacific (8)	1980	11.1	15.8	10.5	5.6	1.4	7.9	47.6
	1990	9.2	9.8	14.5	7.0	1.6	4.1	53.6
	2000	6.9	6.9	16.4	6.2	8.5	2.1	53.1
	2007	6.5	7.2	13.8	4.2	10.2	1.2	57.1
Europe and Central Asia (9)	1980							
	1995	1.4	3.7	2.0	7.0	2.2	8.8	74.9
	2000	2.8	15.3	6.7	4.1	11.2	3.0	56.8
	2007	2.1	9.9	6.4	7.4	8.6	3.4	62.3
Latin America and the Caribbean (10)	1980	6.9	3.6	17.9	4.4	14.4	5.8	47.1
	1990	3.8	5.8	16.3	4.1	3.4	4.4	62.2
	2000	3.9	5.2	23.7	7.8	7.3	3.9	48.0
	2007	1.9	3.3	25.9	19.1	5.8	2.2	41.8
Middle East and North Africa (7)	1980	4.5	17.5	15.6	4.5	8.6	5.1	44.2
	1990	4.9	13.3	18.7	9.0	8.4	4.8	40.9
	2000	4.4	15.1	14.8	10.5	12.7	8.8	33.6
	2007	3.1	10.5	11.8	7.7	24.4	3.5	39.0
South Asia (7)	1980	6.6	19.2	2.9	2.0	4.2	4.3	60.8
	1990	6.9	18.1	3.1	1.8	1.9	3.1	65.0
	2000	4.8	15.3	3.4	1.8	1.8	2.2	70.7
	2007	4.9	12.9	4.6	2.3	1.6	3.2	70.5
Sub-Saharan Africa (10)	1980	6.0	6.1	11.9	3.4	7.8	13.9	50.9
	1990	6.0	8.4	13.9	4.5	3.0	6.0	58.1
	2000	3.6	6.1	15.5	4.7	3.1	3.8	63.3
	2007	2.7	5.4	16.5	7.3	3.5	3.6	61.1

Notes: Calculations include 51 low- and middle-income countries. The number of countries included in each group is shown in parentheses. For countries in Europe and Central Asia estimates are for the years 1995 to 2007. The category "Others" refers to total government spending on all sectors other than the remaining six sectors identified above. Public expenditures on agricultural research and development are included in the "Others" category.
Source: Authors' calculations using IFPRI, 2010.

TABLE 6
Public expenditures on agricultural research and development in 2000, by region

COUNTRY CATEGORY	SPENDING	SHARE
	(Million constant 2005 PPP dollars)	*(Percentage)*
Low- and middle-income countries (131)	**11 441**	**46**
East Asia and Pacific, excluding China (19)	1 192	5
China (1)	1 745	7
Eastern Europe and Former Soviet States (23)	1 177	5
South Asia, excluding India (5)	358	1
India (1)	1 487	6
Latin America and the Caribbean (25)	2 755	11
Sub-Saharan Africa (45)	1 315	5
West Asia and North Africa (12)	1 412	6
High-income countries (40)	**13 456**	**54**
Total (171 countries)	**24 897**	**100**

Note: The number of countries included in each group is shown in parentheses.
Source: IFPRI, 2012a. See Annex table A6.

come at the cost of either increased taxation or a decline in other expenditures, some of which may be socially desirable in their own right and have a significant development impact, including on agricultural productivity and development. It is therefore particularly important to enhance the effectiveness and impact of public expenditures on agriculture, even within existing budget constraints. The allocation of expenditures within agricultural budgets may be more important than overall agricultural expenditure levels (see Chapter 5).

Public expenditures on agricultural research and development

Levels of public expenditure on agricultural research and development

Agricultural research and development (R&D) is a key component of public expenditures on agriculture and is one of the most crucial contributors to agricultural productivity growth. The data on agricultural R&D are reported separately from other agricultural government expenditures. The data do not clearly distinguish between investment and current expenditures, but the literature on returns on spending on agricultural R&D almost universally shows very high returns in terms of agricultural productivity growth and poverty alleviation (see Chapter 5).

According to data compiled by the ASTI initiative managed by IFPRI (2012a), total public expenditures[5] on agricultural R&D worldwide amounted to US$24.9 billion in 2000, the most recent year with complete information (Table 6).[6] Of this, 46 percent was spent by low-and middle-income countries. The 49 low-income countries only accounted for US$2.6 billion, or 10.4 percent.

Public expenditure on agricultural R&D in low- and middle-income countries has increased since 1980 in all regions (Figure 13). The same does not necessarily apply to all countries within the regions (see Annex table A6 for more recent data by country). Indeed, several countries have well-managed and funded systems, producing world-class research; others, some of which are highly dependent on agriculture, have experienced significant reductions in their R&D spending and capacity levels.

[5] Public expenditures include expenditures by governments, institutions of higher education and non-profit organizations.
[6] Data are updated to different years for different regions, but, at the time of writing, 2000 is the most recent year for which complete information is available for all regions. Preliminary results from a global update through 2008 indicate major growth in public spending on agricultural R&D, driven mainly by increases in spending by China and India as well as a number of other large, often more advanced economies.

FIGURE 13
Public expenditures on agricultural research and development, by region

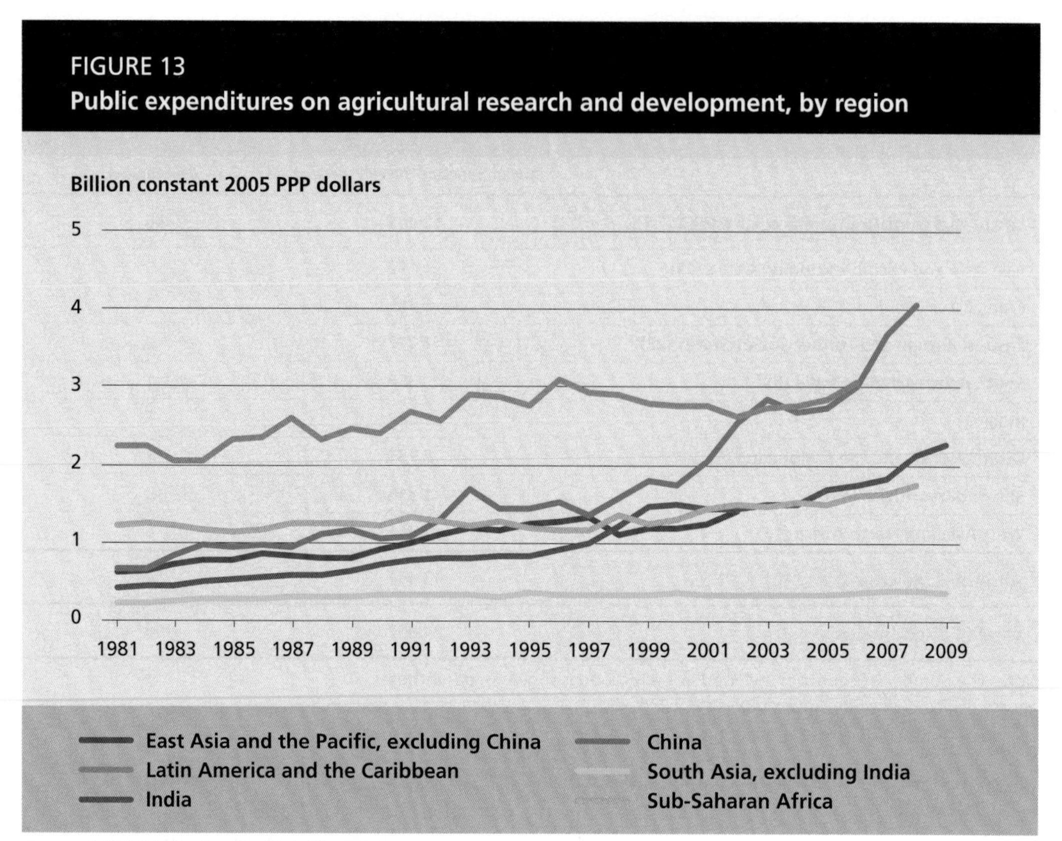

Source: IFPRI, 2012a. See Annex table A6.

In high-income countries, levels of private agricultural R&D are significant, but in the developing world R&D by the private sector remains small (Beintema and Stads, 2008a; Pray, Fuglie and Johnson, 2007; Echeverría and Beintema, 2009). Agricultural R&D in low- and middle-income countries thus depends critically on adequate public funding for these activities.

Most public expenditure on agricultural R&D in the low- and middle-income countries is highly concentrated in a few large countries. China accounted for about two-thirds of total public agricultural R&D spending in East Asia and the Pacific in 2002 (the latest year with available data for the entire region). China's agricultural research spending has continued to expand rapidly ever since. Other countries such as Malaysia and Viet Nam have also realized impressive growth since the early 1990s. In Latin America and the Caribbean, Argentina, Brazil, and Mexico account for the bulk of regional spending, with Brazil alone representing 42 percent of the region's total in 2006. In South Asia, India accounted for 86 percent of the total spending in 2009 (the latest year with available data for the subregion).

In sub-Saharan Africa, after a decade of stagnation in the 1990s, investment in agricultural research in the region rose by more than 20 percent between 2001 and 2008. However, most of this growth occurred in only a few countries. Agricultural research spending in most of the remaining countries in the region, especially in francophone West Africa, has stagnated or fallen since the turn of the millennium.

It is important to assess the magnitude of agricultural R&D efforts relative to the economic significance of the sector. High-income countries spent, on average, 2.4 percent of their agricultural GDP on public agricultural R&D in 2000 (Table 7), while low- and middle-income countries spent significantly less in relative terms (0.5 percent). A target of 1 percent has been recommended by the recent literature as an adequate share for developing countries (Beintema and Elliott, 2011).[7] Considering the significance of private R&D expenditures in high-income countries compared with

[7] As with all indicators, this has several limitations and needs to be considered within the appropriate context (Beintema and Stads, [2008b]).

their limited role in developing countries, the difference between shares in the two groups would be even sharper if private R&D expenditures were included in the comparison.

The lowest regional average is found in South Asia (0.3 percent in 2009) and the highest in Latin America and the Caribbean – the only low- and middle-income region with an average above 1 percent. However, even in this region the ratio is only half that of the high-income countries. Furthermore, large variations at country-level exist within regions (see Annex table A6). Most regions have seen an upward trend in the share of R&D in agricultural GDP. The main exception is sub-Saharan Africa, where the share declined significantly between 1981 and 2000. The downward trend in the region has since been reversed, but the share in the region remains below that of 1981.

Official development assistance to agriculture

Official development assistance (ODA) can contribute to public investment in agriculture, although it is not always clear what share of ODA should be considered investment rather than current expenditure. ODA has been receiving renewed international attention following the food price crisis of 2008. Although overall levels of ODA to agriculture are relatively small compared with government expenditures on agriculture, they may be more significant for individual countries that are major recipients of ODA.

Data from the OECD's creditor reporting system on ODA (Figure 14) indicate that commitments to agriculture peaked in the 1980s – after having grown significantly in the years following the international food crisis of 1973–74 (see Annex able A7 for data by country). During the 1990s, ODA commitments to agriculture decreased continuously, both in absolute terms (measured in constant prices) and as a share of total ODA. Since the mid-2000s, renewed international attention to agricultural development and concerns about rising international food prices have led to partial recovery in the level of ODA to agriculture and its share in total ODA, but both (especially the share) remain well below earlier levels.

New data compiled by FAO with a more comprehensive coverage of donors (FAO, 2012a) show that annual commitments to

TABLE 7
Public expenditures on agricultural research and development as a share of agricultural GDP, by region

COUNTRY CATEGORY	1981	1991	2000	LATEST YEAR
		(Percentage)		
Low- and middle-income countries (108)	**0.55**	**0.54**	**0.54**	..
Sub-Saharan Africa (45)	0.75	0.61	0.55	0.61 (2008)
East Asia and the Pacific, excluding China (19)	0.41	0.51	0.51	0.57 (2002)
China (1)	0.38	0.34	0.38	0.50 (2008)
South Asia, excluding India (5)	0.37	0.39	0.31	0.25 (2009)
India (1)	0.22	0.29	0.39	0.40 (2009)
Latin America and the Caribbean (25)	0.90	1.08	1.21	1.18 (2006)
West Asia and North Africa (12)	0.60	0.59	0.74	..
High-income countries (32)	**1.53**	**2.11**	**2.37**	..
Total (140)	**0.91**	**0.98**	**0.97**	..

Notes: Table excludes 31 countries in Eastern Europe and the former Union of Soviet Socialist Republics, because of data unavailability.
.. = data not available.
Sources: Data on public expenditures on agricultural research and development are from IFPRI (2012a). Data on agricultural GDP are from the World Bank's *World Development Indicators* (2012). See Annex table A6.

BOX 7
Sources of productivity growth in agriculture

There is strong evidence that gains in agricultural productivity have contributed significantly to rising farm incomes and reductions in rural and urban poverty.[1] Above, we discussed the importance of agricultural capital for labour productivity, as measured by agricultural GDP per worker. Such partial productivity indicators are important but do not account for all the factors that contribute to productivity growth. Total factor productivity (TFP) attempts to account for all sources of productivity growth in agriculture. It is an index of measured outputs divided by an aggregate index of measured inputs and physical capital such as land, labour, machinery, livestock, chemical fertilizers and pesticides. Growth in TFP thus represents that part of production growth that is not explained by increased use of these factors but by other things such as technological progress, human capital development, improvements in physical infrastructure and government policies, as well as unmeasured factors such as improvements in input quality or depletion of natural resources (Fischer, Byerlee and Edmeades, 2009).

Fuglie (2010) finds that TFP growth has accounted for an increasing share of agricultural output growth. Figure A shows a breakdown of factors contributing to global agricultural output growth over the past five decades. Machinery, livestock, material inputs (especially fertilizer) and land were key drivers of agricultural growth in the 1960s, 1970s and still in the 1980s. As the contributions of increased use of inputs, physical capital and land declined over time, TFP growth became increasingly prominent and by the 1990s and 2000s was by far the most important factor underlying agricultural growth in a global context. This pattern is also evident in developing regions (Figure B). The only region where this pattern does not hold is sub-Saharan Africa (Figure C). Here new land has been the dominant driver of agricultural growth in the period 1981–2009. TFP became the second most important factor in the 1980s, but its contribution has declined over the years, in contrast with that of developing countries as a whole. For sub-Saharan Africa, the transition to sustainable agricultural intensification will require a change from a strategy based on area expansion to one based on investment in activities that enhance TFP growth.

Earlier work by Evenson and Fuglie (2009) examined the relationship between long-run TFP growth and national investment in technology capital for 87 developing countries. They considered both an indicator of the capacity to *develop or adapt* new technology and an indicator of the capacity to *extend and adopt* agricultural technology. They found that rising TFP growth rates were positively correlated with increases in either indicator provided that a minimum capacity existed in the other. Both research and extension were thus found to be important drivers of TFP growth. However, the results pointed to the need to place more emphasis on research relative to extension. Improvements to research capacity were often associated with increased productivity growth even in the absence of improved extension capacity, while the reverse was not true. The results were confirmed in subsequent analysis by Fuglie (2012).

[1] For a sample of the numerous studies on the contribution of agricultural productivity to growth and poverty reduction see Thorbecke and Jung (1996); Datt and Ravallion (1998); Foster and Rosenzweig (2004); Mundlak, Larson and Butzer (2004); Ravallion and Chen (2004); Christiaensen and Demery (2007); Bezemer and Headey (2008); Otsuka, Estudillo and Sawada (2009); and Suryahadi, Suryadarma and Sumarto (2009).

Growth in global agricultural output, by source of growth and time period

A - Global agricultural output

Percentage change

3
2
1
0

1961–2009 1961–1970 1971–1980 1981–1990 1991–2000 2001–2009

B - Developing countries

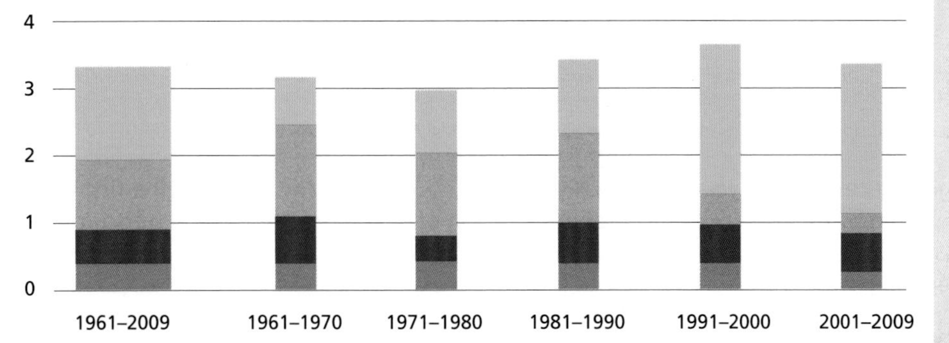

C - Sub-Saharan Africa

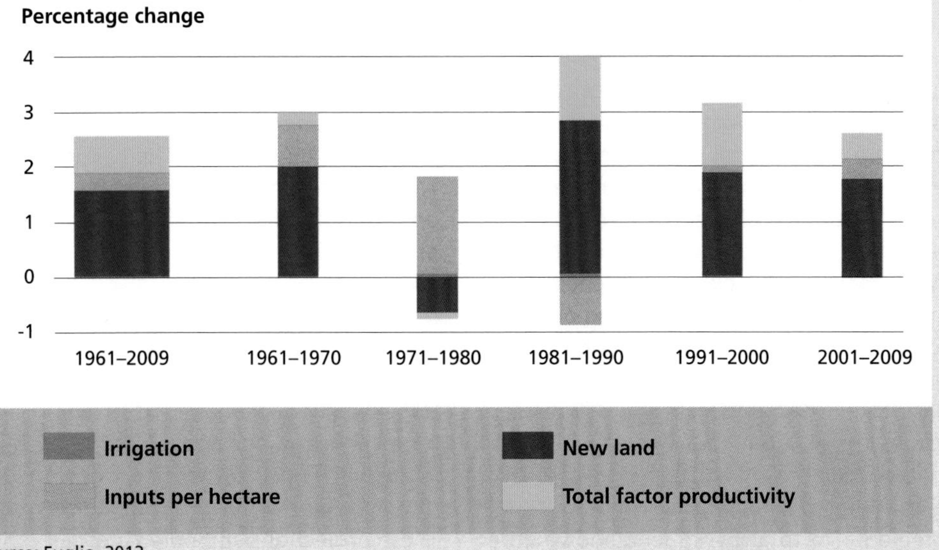

Source: Fuglie, 2012.

FIGURE 14
Level and share of official development assistance committed to agriculture, by region

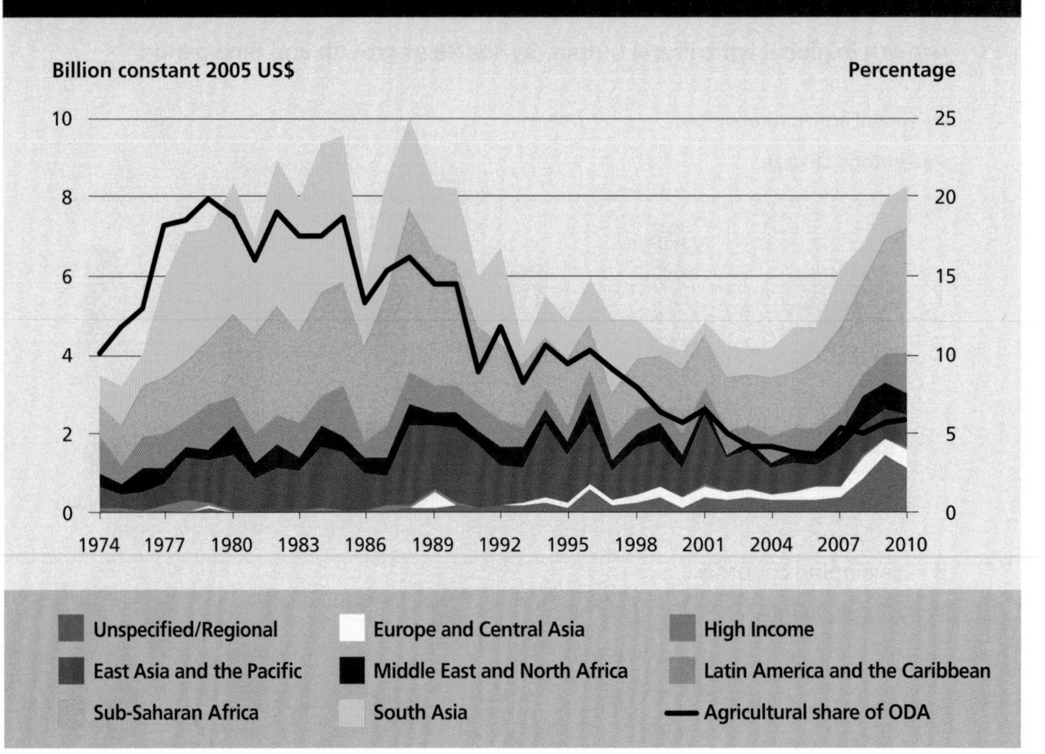

Source: Authors' calculations using data from OECD, 2012a. See Annex table A7.

agriculture in recent years exceeded those reported by the OECD's creditor reporting system by 1–2 billion US$, but confirm the general pattern revealed by the OECD data.

Increasing investment in agriculture

The evidence presented in this chapter suggests that many low- and middle-income countries need to invest more in agriculture. However, assessing exactly how much and what type of additional investment is needed and by whom these investments should be made is more difficult. Several efforts have been made over time – by FAO and others – to estimate overall investment needs in agriculture. These differ, depending on factors such as the specified objective, the time horizon, the sectoral coverage (only primary agriculture or also upstream and downstream sectors), the geographical coverage, whether both private and public investment are considered, whether they consider incremental or total investment, and whether they represent gross or net investment.

As noted in Chapter 1, the first edition of *The State of Food and Agriculture* in 1947 called for increased investment in agriculture to transform less-populated regions in Latin America and Africa into "granaries" for the rest of the world. In 1949, the third edition of *The State of Food and Agriculture* indicated that the low-income countries needed additional foreign capital for investment in support of agriculture of US$4 billion per year to supplement the US$13 billion that they would need to raise themselves (FAO, 1949). The two most recent key global estimates prepared by FAO, based on different objectives and assumptions, are presented in the following.

Meeting demand for food in 2050

In 2009, FAO estimated that average annual investment flows amounting to US$209 billion were needed to meet projected demand for agricultural products in 2050 in 93 developing countries (Schmidhuber, Bruinsma and Bödeker, 2009). These projections embodied a broad range of capital items in primary crop and livestock

BOX 8
The L'Aquila Food Security Initiative

Since the food price crisis of 2008, issues of food security have moved to the forefront of the international agenda. The G8 meeting in L'Aquila, Italy, in July 2009 resulted in a Joint Statement on Global Food Security, which recognized consistent underinvestment in agriculture combined with economic instability as partial reasons for the persistence of food insecurity. It noted the decreasing levels of ODA to agriculture and the need to reverse the trend. The G8 member nations reaffirmed their commitment to improve food security and pledged US$20 billion in assistance to agriculture and food security in developing countries over the following three years (G8, 2009). At a meeting of the G20 in Pittsburgh in September 2009, the amount was increased to US$22 billion and the Global Agriculture and Food Security Program (GAFSP) was established to assist in delivery on the pledges.

The GAFSP is housed at the World Bank and is governed by a Steering Committee with wide representation by major donor and recipient countries and international organizations, including the multilateral development banks, IFAD, FAO, WFP, the International Finance Corporation (IFC) and the UN Secretariat. It aims to increase both the level and predictability of ODA to agriculture, by reviewing proposals by donors and by monitoring and evaluating project implementation. From its inception through February 2012 the GAFSP had approved proposals for projects totalling 1.1 billion US$ to be implemented in Cambodia, Ethiopia, Haiti, Liberia, Mongolia, Nepal, the Niger, Rwanda, Sierra Leone, Tajikistan and Togo.

The L'Aquila initiative has been criticized for failing to specify whether the pledged funds were additional to existing levels of ODA or to provide clear definitions of what was meant by aid, agriculture and food security. While there is no official monitoring of delivery on the L'Aquila pledges, FAO, in response to recommendations by the renewed Committee on World Food Security (CFS), has developed the Mapping Actions for Food Security and Nutrition web-based platform, which allows countries to track and map their investment in support of food security and nutrition. (FAO, 2011c). Despite the L'Aquila pledges, ODA commitments to agriculture increased only about one-third of a billion US$ from 2009 to 2010 (OECD, 2012a).

production as well as downstream support services,[8] and they were made under specific assumptions regarding key parameters such as population growth and urbanization. Of the total, US$83 billion represent net investment, with the residual corresponding to the cost of replacing depreciating capital. A breakdown of the average annual investment needs from 2005–07 to 2050 by region and aggregate type of investment is shown in Figure 15.

These estimates represent the level of investment required to meet growing demand for food in 2050 – not to eliminate hunger, although they do imply some reduction in poverty and hunger. Specifically targeting poverty or undernourishment implies assessing how much more investment is needed in addition to these projections or to some other "business as usual" scenario.

Targeting poverty and hunger

In a separate analysis, Schmidhuber and Bruinsma (2011) provide estimates of incremental public expenditures on agriculture and safety nets needed to reach a world free of hunger by 2025. Over this period, incremental annual public expenditures of US$50.2 billion

[8] Main categories included are as follows. For crop production: land development, soil conservation and flood control, expansion and improvement of irrigation, permanent crop establishment, mechanization, other power sources and equipment, working capital. For livestock production: herd increases, meat and milk production. For downstream support services: cold and dry storage, rural and wholesale market facilities, first-stage processing. No distinction is made regarding whether investments will be financed from private or public sources.

FIGURE 15
Average annual investment needs in low- and middle-income countries, by region

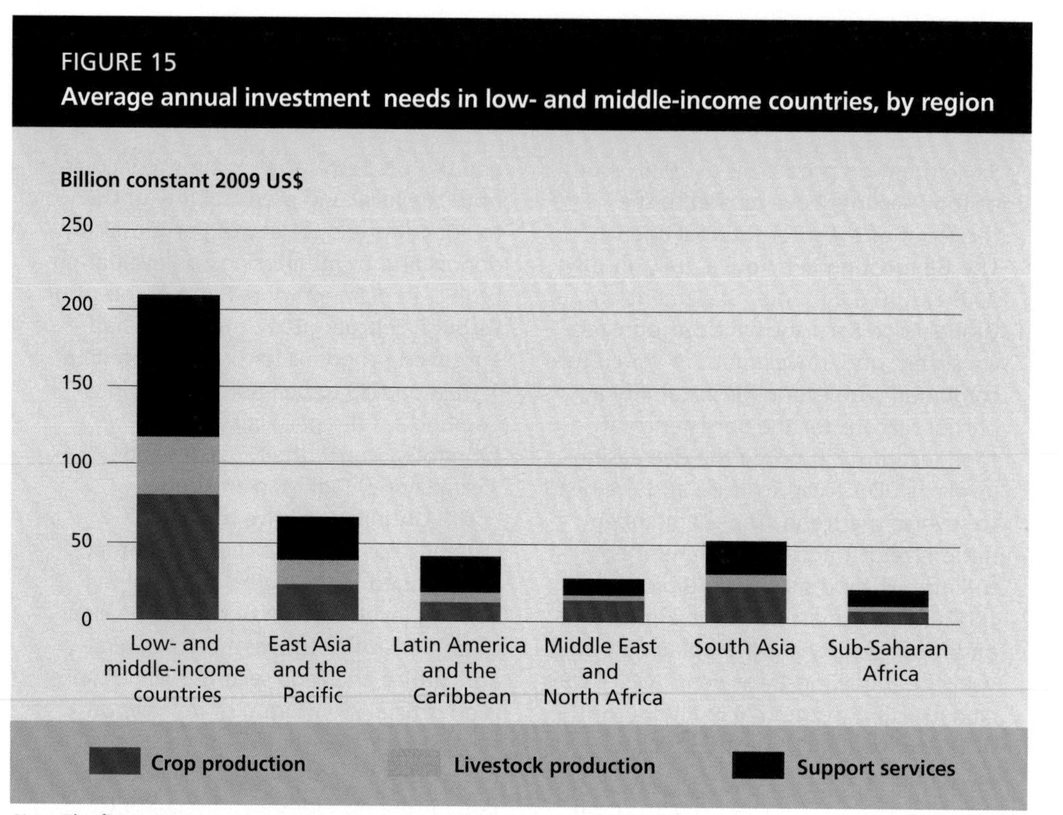

Note: The figure presents average annual needs over the period 2005–07 to 2050.
Source: Schmidhuber, Bruinsma and Bödeker, 2009.

TABLE 8
Incremental annual public investment needed to eradicate hunger by 2025

PRIORITY AREA FOR INVESTMENT	INVESTMENT NEEDED
	(Billion constant 2009 US$)
1. Expand rural infrastructure and market access	18.5
2. Develop and conserve natural resources	9.4
3. Research, development and extension	6.3
4. Rural institutions	5.6
5. Expenditures for safety nets	10.4
Total investment costs	**50.2**

Source: Schmidhuber and Bruinsma, 2011.

are estimated to be required (in addition to existing levels of spending) to support investment in rural infrastructure, natural resource conservation, research, development and extension, and rural institutions, but also to provide safety nets aimed at those suffering from hunger (Table 8).

Making the transition to sustainability

Meeting future demand growth sustainably, while accelerating the reduction of poverty and hunger, will require even higher levels of investment by farmers and the public sector. Analysis of sustainable production systems often shows them to be beneficial in terms of both increasing returns to producers and improving the environment (Pretty *et al.*, 2006). Yet the relatively low adoption rate of such systems seems to indicate they are not attractive to producers.

Moving to sustainable production systems involves significant immediate costs, not only in the form of investment and operating expenses, but also opportunity costs – for example the income producers forego during the transition to a new system. It can be several years before positive returns to sustainable agricultural systems are realized, particularly where they involve restoration of degraded ecosystems (McCarthy, Lipper and Branca, 2011). Few producers can finance such a long period of lost income – even if they stand to make major gains in the future (see also Box 14). Transaction costs can also be an obstacle to adopting sustainable practices. Sustainable production systems require more coordination, for example in managing common-property natural resources, or in coordinating post-harvest, processing, storage and marketing activities. This implies significant investments in social capital. Transitioning to sustainable consumption systems incurs a similar set of costs. Reducing waste involves not only investment and operating costs, but also the transaction costs of coordination among production, processing, storage and marketing phases.

Several governments in low- and middle-income countries have begun supporting farmers in the transition to more sustainable production practices. For example, the Government of Zambia adopted conservation agriculture as a policy priority in late 1999 in order to improve agricultural productivity and sustainability. It created the Conservation Farming Unit, which now provides extension services to 170 000 farmers in 17 districts to support the adoption of conservation agriculture. The technology has been most successful in semi-arid regions because it reduces the effects of drought on agricultural productivity without sacrificing yields. Even in these regions, however, many farmers have abandoned the practice, suggesting that more needs to be known about the institutional, agro-ecological and economic factors that influence the successful adoption of more sustainable agricultural practices (Arslan *et al.*, 2012). Similarly, the Government of Malawi supported the establishment of a National Task Force on conservation agriculture in 2002 and reports that 18 471 hectares, 110 percent of the target, are cultivated using conservation agriculture (Malawi Ministry of Agriculture, Irrigation and Water Development, 2012). The Government of Viet Nam has also embraced sustainable development of agricultural production, especially sustainable rice intensification, which has significant potential in improving food security and decreasing greenhouse gas emissions, while improving farmers' capacities to adapt to the effects of climate change.

Appropriate institutions and policies can reduce the costs individual investors face in moving to sustainable systems. For example, social safety nets and programmes to reduce risk and strengthen resilience *ex-ante* can strengthen incentives for investments in sustainable systems (FAO, 2010a). Publicly provided agricultural research, development and extension systems, combined with capacity building, reduce transaction costs and increase incentives for investments in sustainable practices. The reallocation of existing public and private investment resources – moving from investments that have low "sustainability" returns to higher ones – is key to moving towards sustainable production systems. Ensuring that environmental goods and services are incorporated into investment incentives is a crucial policy challenge (see Chapter 3). Similarly, agricultural research and development is essential for underpinning sustainable approaches in agriculture.

Potential new and additional sources of financing that could channel more private-sector finance towards sustainable development include payments for the provision of environmental public goods (such as biodiversity conservation, climate change mitigation or protection of water bodies). Linking climate change finance to sustainable agricultural investment plans could also provide additional finance (both are discussed further in Chapter 3).

The challenge of fostering investments in agriculture

The relative magnitude of investment flows from public and private sources clearly shows that private investment is the key to meeting future demand growth, achieving food security and making the transition to sustainable agriculture. But governments can only facilitate private investment by farmers and other investors. The question facing policy-makers therefore is "What is required to ensure that adequate agricultural investments occur and that they meet the objectives of food security, poverty alleviation and environmental sustainability?" This question will be addressed in the following chapters.

Key messages

- Private investment by farmers themselves is the largest source of investment in agriculture in low- and middle-income countries, far exceeding the annual flows to agriculture from governments, donors and foreign investors. The roles of public and private investors are complementary and generally cannot be substituted for each other, but the central role of farmers must be recognized in any strategy that seeks to promote agricultural investment.
- Systematic and comprehensive data on agricultural investment are very limited. A few internationally comparable datasets shed some light on different aspects of investments in agriculture, but improved data are necessary to clarify the levels and trends in agricultural investment and to enable more robust analysis of the impacts of different types of investment.
- Agricultural capital stock – especially agricultural capital stock per worker – is an important determinant of agricultural labour productivity. There are large gaps in agricultural capital–labour ratios between the high-income countries and the middle- and low-income countries. The gap between high-income and low-income countries has widened over recent decades as agricultural capital stock in the low-income countries has been outpaced by growth in the labour force. In particular sub-Saharan Africa and South Asia has seen declining and stagnant capital–labour ratios during this period.
- FDI in agriculture has increased in recent years but it represents a very small portion of total FDI flows and of total resource flows to agriculture in low-and middle-income countries. FDI is unlikely to make a significant contribution at the global level to increasing agricultural capital stock per worker, but it is a major factor for some individual countries.
- Public investment in agriculture is necessary to promote private investment in the sector, but governments in low- and middle-income countries have devoted a declining share of public expenditures to agriculture. The regions with the highest incidence of undernourishment – sub-Saharan Africa and South Asia – are also the ones that devote the smallest share of expenditure to agriculture relative to agriculture's share in GDP.
- Overall, low- and middle-income countries spend significantly less on R&D as a share of agricultural GDP than the high-income countries, and most of these expenditures are concentrated in relatively few countries. Given the positive role of R&D in promoting agricultural growth and poverty reduction, there is an urgent need to increase R&D funding for agriculture in the low- and middle-income countries.
- Globally, flows of ODA comprise a relatively minor share of agricultural investment but can be significant for some countries. After years of continuous decline, in recent years ODA to agriculture has increased both absolutely and as a share of total ODA, while still remaining below the levels of the 1980s.
- The relative importance of private investment means that the investment climate in which farmers make decisions is critical. It is the responsibility of governments to create the conditions to foster investment in agriculture.

3. Fostering farmers' investment in agriculture

Most investment in agriculture is made on the half a billion farms located around the world.[9] On-farm agricultural investment decisions are based on the potential profitability and risks compared with other investment opportunities and the individual constraints they face. In any country, the relative returns, risks and constraints associated with agricultural investment are affected by the overall investment climate, policies specific to agriculture and the provision of public goods that are essential for agriculture. Governments of countries that are dependent on agriculture for a large share of employment and GDP have a responsibility to provide an investment climate that is conducive to investment in the sector. Ensuring that agriculture is not penalized relative to other sectors is a basic element of this. Along with the need to foster investment in agriculture, governments have a responsibility to ensure that such investment is environmentally sustainable.

This chapter reviews the issues involved in creating a climate that fosters sustainable investment by farmers. It first looks at the role of the overall investment climate in promoting agricultural investment. It then discusses more specifically the role of economic incentives to invest in agriculture and how they are shaped by policies in agriculture and other sectors. It also discusses the crucial role of agro-industries in transmitting price incentives to farmers and briefly considers the enabling conditions for investment in these industries. The chapter concludes with a discussion of how to ensure that environmental costs and benefits are appropriately included in incentives to invest in agriculture in order to promote sustainability and socially beneficial outcomes.

[9] Nagayets (2005) estimated the total number of farms in the world to approximately 525 million.

Creating a favourable climate for investment in agriculture

Considerable attention has been focused on what constitutes a proper climate for private investment. Less has been given to how important these factors are for investment in agriculture. The *World Development Report 2005* argued that the general investment climate is central to growth and poverty reduction:

> *The investment climate reflects the many location-specific factors that shape the opportunities and incentives for firms to invest productively, create jobs, and expand. A good investment climate is not just about generating profits for firms – if that were the goal, the focus could be limited to minimizing costs and risks. A good investment climate improves outcomes for society as a whole.*
>
> (World Bank, 2004, p. 2).

According to the World Bank, the roles of government in providing a good general investment climate include:

- ensuring stability and security, including rights to land and other property, contract enforcement and crime reduction;
- improving regulation and taxation, both domestically and at the border;
- providing infrastructure and financial market institutions; and
- facilitating labour markets by fostering a skilled workforce, crafting flexible and fair labour regulation and helping workers cope with change.

Each of these elements is complex and location-specific. Several indicators have been developed by the World Bank and other international organizations and research institutions to assess the business and investment climate in different countries.

FIGURE 16
Worldwide Governance Indicator for Rule of Law and agricultural capital stock per worker, by country

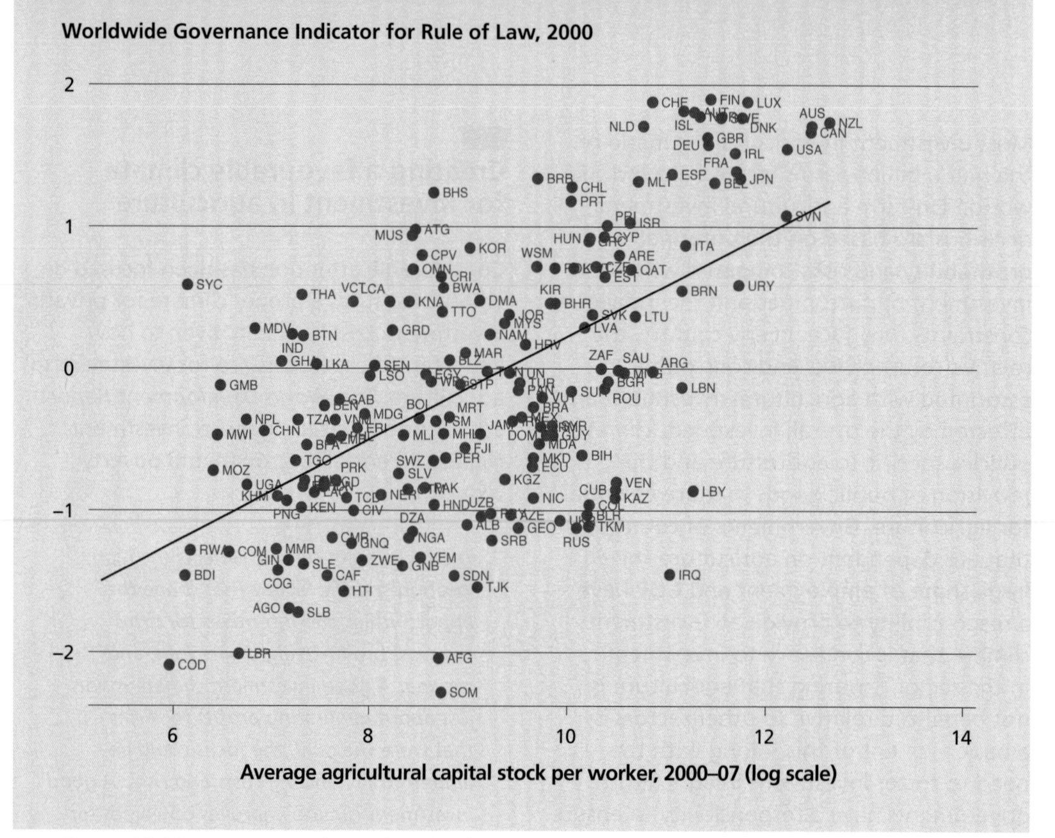

Note: The Worldwide Governance Indicator for Rule of Law ranges from −2 to 2 with smaller values indicating poorer rule of law.
Source: Authors' compilation using World Bank, 2011c and FAO, 2012a. See Annex table A2.

Some indicators provide an assessment of overall governance in a country, including dimensions such as democracy, rule of law, absence of conflicts, and corruption. Others deal more specifically with factors that affect the ease of doing business in a country.

Despite the economic importance of agriculture in most low- and middle-income countries, insufficient attention has been given to assessing the extent to which these elements of a good investment climate are relevant for agriculture. Analysis of the relationship between these indicators and agricultural capital stock undertaken for this report suggests that they are indeed highly relevant.

Governance and agricultural investment

A commonly used indicator of the governance in a country, the Worldwide Governance Indicator for Rule of Law,[10] is closely correlated with agricultural capital stock per worker (Figure 16). Comparable patterns also emerge for other governance indicators, such as the Corruption Perception Index[11] compiled by Transparency International and the Political Risk Index[12] of the Political Risk Services

[10] The Worldwide Governance Indicator for Rule of Law measures overall crime rates and the extent to which agents believe in and follow laws, especially those pertaining to contract enforcement, property rights and the court systems.

[11] The Corruption Perception Index measures public-sector corruption in the country and covers, *inter alia*, the embezzlement of public funds, bribery of officials and the effectiveness of anti-corruption measures.

[12] The Political Risk Index measures government stability, socio-economic conditions, risk associated with investments, internal and external conflicts, corruption, degree of influence of military and religion in politics, law and order, ethnic tensions, democratic accountability and quality of bureaucracy.

TABLE 9
Business environment rankings and on farm investment in low- and middle-income countries

"DOING BUSINESS" RANKING	AGRICULTURAL CAPITAL STOCK PER WORKER, 2007	ANNUAL CHANGE IN AGRICULTURAL CAPITAL STOCK PER WORKER, 1995–2007
	(Constant 2005 US$)	*(Percentage)*
Top ten (best business environment)	19 000	2.4
Bottom ten (worst business environment)	5 600	0.3

Source: Authors' compilation based on World Bank, 2011d and FAO, 2012a. See Annex table A2.

Group. While these relationships only show correlations, they strongly suggest that the same elements of good governance that are needed for overall investment in an economy are equally needed for agriculture. Further evidence presented in Chapter 5 supports the conclusion that arbitrary, corrupt and unstable governments are not conducive to agricultural investment.

The investment climate and agricultural investment

Beyond governance, other factors may directly facilitate or impede the operations of economic agents or investors in a country, such as access to transport, finance and electricity. These factors are difficult to quantify, and data are scarce, but interesting patterns emerge from World Bank indicators describing the urban business climate, including rankings of the Ease of Doing Business,[13] based on interviews with experts on private-sector activities in the various countries. The rankings show a clear relationship with the level and growth of agricultural capital stock per worker (Table 9). Looking only at the low- and middle-income countries, the ten countries where it is easiest to do business had more than three times the agricultural capital per worker in 2007 as the ten most challenging countries (US$19 000 versus US$5 600). Moreover, the rate of growth in agricultural capital stock per worker since 2000 was eight times faster in the most favourable countries compared with the least favourable. Again, while these are only correlations, they strongly suggest that the factors that facilitate general business investment are likewise important for agriculture.

Rural Investment Climate Assessments

The rankings discussed above are derived from urban settings, but efforts are under way to compile indicators for the rural investment climate.[14] The Rural Investment Climate Assessments by the World Bank – conducted for a small number of countries – examine small and medium off-farm enterprises located in rural areas. These assessments do not include primary agriculture, but a favourable investment climate for small enterprises in rural areas would be expected to have beneficial impacts on agriculture also, not least because of the importance of agro-processing and marketing enterprises in influencing incentives for agricultural investment.

The Rural Investment Climate Assessments found that the constraints faced by rural enterprises differ from those faced by large urban firms and vary by location; thus the priorities for governments and donors interested in addressing these constraints must be context-specific. For example, Table 10 shows the top five constraints cited by large firms in urban areas versus those identified by small and medium enterprises in rural areas (and the share of firms citing them as such) in Nicaragua, Sri Lanka and the United Republic of Tanzania. Transport appears more problematic for rural enterprises than urban ones in Sri Lanka and the United Republic of Tanzania, and lack of demand or marketing constraints pose more difficulties in rural areas than in urban areas in all three countries.

[13] The Ease of Doing Business rankings measure how easy it is to open and close a business, deal with construction permits, register property, obtain credit, pay taxes and trade across borders as well as how effectively investments are protected and contracts enforced.

[14] Agribusiness indicators are being developed by the World Bank's Agriculture Department, and the Bill and Melinda Gates Foundation is developing indicators of Doing Business in Agriculture.

TABLE 10
Top five greatest constraints to business activities identified by urban versus rural firms in selected countries

SRI LANKA			
Factor	Percentage of urban firms	Factor	Percentage of rural firms
Electricity	42	Transport	30
Policy uncertainty	35	Finance (cost of)	28
Macroeconomic instability	28	Finance (access)	28
Finance (cost of)	27	Demand	27
Labour regulation	25	Electricity	26
UNITED REPUBLIC OF TANZANIA			
Factor	Percentage of urban firms	Factor	Percentage of rural firms
Tax rate	73	Finance	61
Electricity	59	Utilities (electricity)	49
Finance (cost of)	58	Transport	30
Tax administration	56	Marketing	29
Corruption	51	Governance	27
NICARAGUA			
Factor	Percentage of urban firms	Factor	Percentage of rural firms
Corruption	65	Political uncertainty	53
Finance (cost of)	58	Electricity	41
Economic/regulatory uncertainty	56	Corruption	39
Finance (access)	54	Finance (access)	38
Finance (availability)	49	Lack of demand	30

Note: Firms surveyed in urban areas were large-scale whereas those in rural areas were small and medium off-farm enterprises.
Source: World Bank, 2006b.

Some factors, including the provision of public services such as electricity and concerns about corruption, political instability and governance, are key constraints cited by investors in both rural and urban settings.

An enabling environment for agricultural value chains

Agricultural value chains include many enterprises that provide goods and services to farmers, such as input supplies, storage and processing facilities and marketing services. They provide a crucial link between farmers and markets, upstream and downstream, and are essential for the effective transmission of investment incentives from markets to farmers (da Silva *et al.*, 2009).

As most developing countries become increasingly urbanized, so the distance between farmers and consumers is becoming greater. The role of agro-industries is becoming ever more prominent in mediating demand for food to primary producers. The development of these sectors can significantly improve the returns to agricultural production and incentives for investment by farmers. Such development extends beyond large-scale enterprises to, especially, small and medium-sized enterprises, which may link better with smallholders (de Janvry, 2009).

An enabling environment for agro-industry development can encourage the entry of small- and medium-sized enterprises into the market and foster their competitiveness. The overall investment climate, as well as the specific rural investment climate discussed above, is critical for such development. Christy *et al.* (2009) examined in more detail the key

BOX 9
An enabling environment for agro-industries

Creating a favourable policy environment for agro-industry can provide a significant contribution to generating investment in primary agriculture. Christy *et al.* (2009) argue the case for a specialized method of describing the competitive environment for agribusiness firms. They examine necessary state actions required to create an enabling environment for competitive agro-industries and propose a hierarchy of essential, important and useful "enablers" that influence agro-industry competitiveness. The pyramid below shows a hierarchy of these enabling needs.

Hierarchy of enabling needs for agro-industry competitiveness

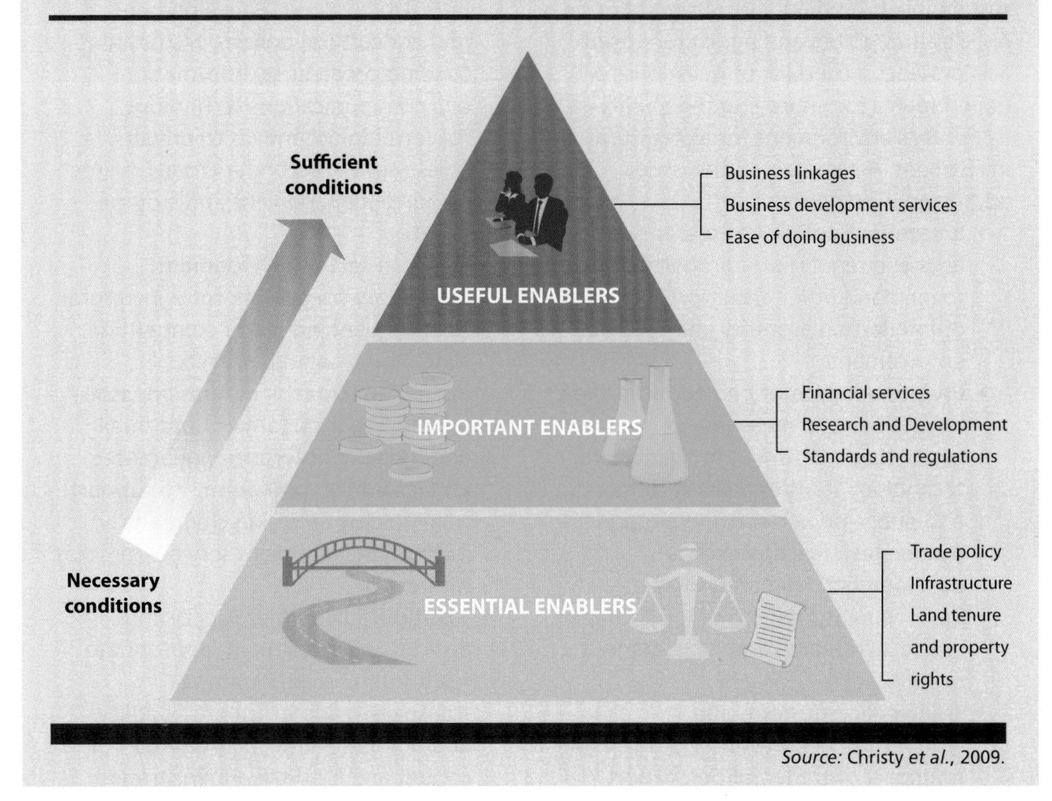

Source: Christy *et al.*, 2009.

components of an enabling environment for competitive agro-industries (Box 9).

A policy framework for agricultural investment

Having recognized the importance of an enabling environment for agricultural investment, the OECD and the New Partnership for Africa's Development (NEPAD) have developed a draft policy framework for promoting investment in the sector (Box 10). It underlines the complexity and diversity of the issues involved in ensuring an appropriate environment for agricultural investment and the extent to which the necessary policies and institutions to a large extent transcend agriculture. They emphasize many of the well-known elements of good governance for investment identified above.

Government policies and incentives to invest in agriculture[15]

Government policies and market interventions can have a profound impact on the investment climate for agriculture and, specifically, on the

[15] This section draws on a background paper prepared by Kym Anderson (University of Adelaide) for the *State of Food and Agriculture 2012*.

BOX 10
The NEPAD-OECD draft Policy Framework for Investment in Agriculture

The draft *Policy Framework for Investment in Agriculture* was prepared within the framework of the NEPAD-OECD Africa Investment Initiative and presented at the 5th NEPAD-OECD Ministerial Conference, held on 26–27 April 2011. It is intended as a flexible tool for governments to evaluate and design policies for agricultural investment in Africa. It recognizes that sustainable growth in agriculture relies on policies that go beyond agriculture itself and provides a checklist of questions for governments for improving the quality of a country's environment for agricultural investment. A summary of the issues addressed is as follows:

- **Investment policy.** Transparency of laws and regulations, property rights to land and other assets, protection of intellectual property and contract enforcement.
- **Investment promotion and facilitation.** Institutions and measures for promoting investment in agriculture, technology transfer to local farmers and public-private sector dialogue.
- **Human resource and skills development.** Human resource development, training of local farmers and local research and development capacity.
- **Trade policy.** Customs and administrative procedures, assessment of impact of trade policies, export promotion and financing, regional trade agreements.
- **Environment.** Policies for natural resource management and cleaner technologies, integration of R&D and environmental policies, energy needs and mitigation of extreme weather.
- **Responsible business conduct.** Labour standards in agriculture, enforcement of human rights, environmental protection, labour relations and financial accountability.
- **Infrastructure development.** Coherent infrastructure, rural development and agricultural policies, transparent funding procedures, information and communications technology for farming, incentives to private investment in secondary roads, water resource management and storage facilities.
- **Financial sector development.** Regulatory framework for agricultural finance, banking sector competition, functioning capital markets, instruments for risk mitigation, access to credit by local farmers and small and medium enterprises, guarantee and insurance mechanisms to support smallholders accessing credit and business development services for local farmers.
- **Taxation.** Tax policies supporting agricultural investment, appropriate tax burden on agribusiness, transparent and efficient tax policy and administration, coordination of central and local tax administration and funding of local public goods.

Source: OECD, 2011.

economic incentives to invest in the sector. Some of these are specific to agriculture, but others relate to other sectors or are economy-wide (Schiff and Valdés, 2002). The main sector-specific policies affecting incentives in agriculture include tariffs, input and credit subsidies, price controls, quantitative trade restrictions, government expenditures and taxes. There may also be indirect effects on agriculture deriving from other policies, such as protection of other sectors (e.g. industrial protection), exchange and interest rates, fiscal and monetary policies. Such policies may significantly affect the incentives to invest in agriculture relative to other sectors.

In the mid-twentieth century, many developing countries implemented policies aimed at promoting industrial development. These policies created a bias against agriculture and disincentives for investment and production. In many developed

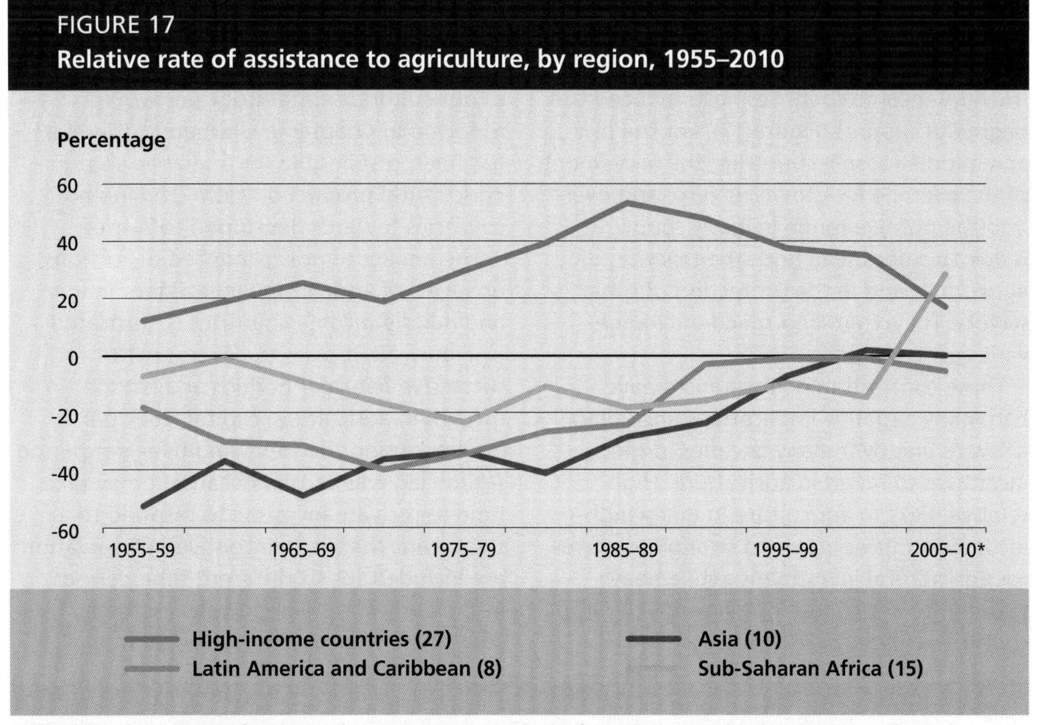

FIGURE 17
Relative rate of assistance to agriculture, by region, 1955–2010

Notes: * For the majority of countries the most recent point in the figure is the weighted average annual rate (based on agricultural production) for the years 2005–10; otherwise it is the value for the most recent observations. For all other periods the values are five-year weighted annual averages. The number of countries included in each group is shown in parentheses.
Source: Authors' calculations using data from Anderson and Nelgen, 2012.

countries, on the other hand, governments wanted to protect the agriculture sector, which was declining in relative economic size but remained socially and politically powerful. These broad trends severely curtailed agriculture in developing countries relative to other sectors at home and relative to developed-country agriculture on world markets. These policy-induced "distortions" of agricultural incentives in developing countries and their associated costs for 18 countries were documented by Krueger, Schiff and Valdés (1988; 1991).

Policy trends and incentives to invest

Over time, policy reforms have changed the levels of protection and taxation of agriculture in many countries and have realigned investment incentives, improving them in developing countries and reducing them in developed countries. These broad trends are presented in recent work by the World Bank on more than 70 countries in developing and developed regions over the past five decades (Anderson and Valenzuela, 2008; Anderson 2009; Anderson and Nelgen, 2012).

An overall indicator of policy-induced price distortions to agriculture, the relative rate of assistance (RRA), measures the extent to which government policies affect farm prices *relative* to other sectors and provides an indication of the degree to which a country's overall policy regime is biased for or against agriculture. A positive RRA implies that agriculture is favoured or subsidized relative to other sectors, while a negative RRA indicates that agriculture is penalized or taxed (Anderson and Valenzuela, 2008). Analysis of average RRAs over time shows the dramatic differences in the policy stance towards agriculture between developed and developing regions. From the mid-1950s, agriculture was taxed heavily in many of the low- and middle-income countries of Asia, Latin America and sub-Saharan Africa, while the sector was increasingly protected in many of the high-income countries (Figure 17).[16]

[16] Australia and New Zealand are exceptional in that they had an anti-agricultural policy bias for most of the twentieth century because their manufacturing tariff protection exceeded agricultural supports. Both sectors' distortions were reduced in the final third of that century and are now close to zero. See Anderson, Lloyd and MacLaren (2007).

Beginning at different times, the low- and middle-income countries have gradually reduced the bias against agriculture and the high-income countries have reduced the degree of support (Figure 17). For the low- and middle-income countries, the movement of the average RRA towards zero, and even into the positive range in many countries, is due to a decline in both the taxation of agriculture and in the protection of other sectors. The extent and speed of change varies across regions.

These contrasting policy stances have had many negative implications, including a severe bias over many decades in the incentives to invest in agriculture. High relative RRAs to agriculture in many high-income countries provided strong incentives for agricultural investment, while heavy taxation of agriculture in many developing countries created severe disincentives. This distorted the geographical pattern of agricultural investment and is partially responsible for the divergence in the levels of agricultural capital stock per worker reported in Chapter 2. Ultimately, this bias has been responsible for a relative shift in agricultural production from developing countries towards developed countries.

The impact of policy-induced distortions on levels of agricultural investment in low- and middle-income countries is illustrated by Figure 18. It plots the average RRA for successive five-year periods against the growth in agricultural capital stock per worker during the following five-year period. A time lag was chosen because it may take time for private investors to respond to a significant change in incentives. Observations are included for six different time periods. In the earliest time period the extreme bias against agriculture, reflected in an

FIGURE 18
Relative rate of assistance and change in agricultural capital stock per worker in low- and middle-income countries

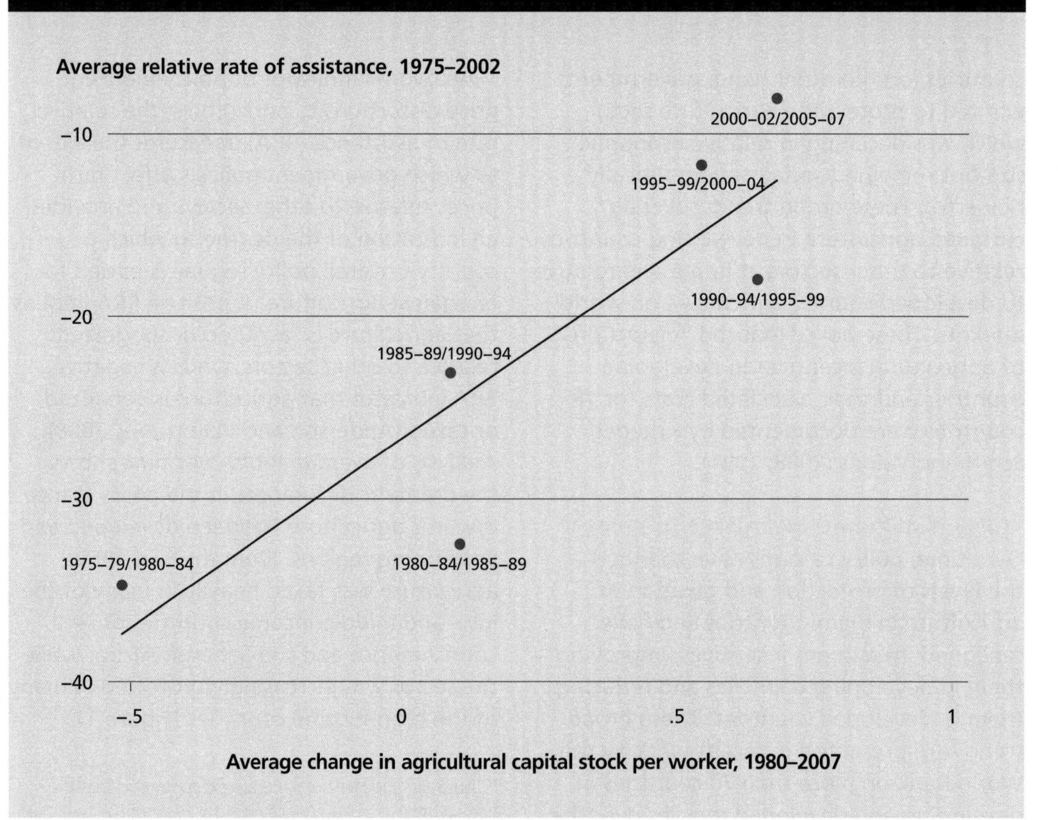

Notes: The indicators are shown for five-year averages. The average relative rate of assistance for each period is plotted against the average annual change in agricultural capital stock per worker in the subsequent five-year period.
Source: Authors' calculations using Anderson and Nelgen, 2012 and FAO, 2012a.

average RRA of around negative 35 percent in 1975–79, is associated with an average annual reduction in agricultural capital stock per worker of more than 0.5 percent in the period 1980–84. Over time, as the average RRA moved from strongly negative closer to neutral, the growth rate for capital stock per worker increased.

Country-level policy distortions

The averages shown in Figure 18 hide considerable variation among RRAs within regions and income groups. Policy-induced price distortions remain significant in many countries, and low- and middle-income countries are still more likely to tax and high-income countries to subsidize agriculture (Figure 19). Some individual country examples are quite extreme. This suggests that efforts to increase agricultural investment in low- and middle-income countries are still being hindered by policy distortions at both extremes of the spectrum.

Commodity-level incentives: focus on Africa

The overall level of protection or taxation of the agriculture sector is clearly important, but differences across commodities within a country also influence investment incentives. There can be considerable variation, with some commodities receiving protection and others being taxed. This may lead to inefficient patterns of investment and production within the country's agriculture sector.

Such differences across commodities may also have different impacts on different types of producers, with some farmers being advantaged over others. This may imply, for instance, different incentives to producers of cash or export crops versus food staples or to smallholders versus large-scale farmers. The impact on a specific commodity is often the result of the interaction of different policies and policy instruments; these can sometimes be inconsistent, with their individual impacts acting in opposite directions. As a result, the incentives for agricultural investment may be unclear to investors and not aligned with the goals of policy-makers.

An improved incentive framework for agriculture requires careful analysis of agricultural policies within a country in order to improve the consistency and transparency of their impacts. The Monitoring African Food and Agricultural Policies (MAFAP) project aims to improve the evidence base for policy-making in ten African countries by providing a framework for analysing the impact of policies and market development gaps (Box 11).

Preliminary results from MAFAP are available for several specific commodities. For example, Figure 20 shows trends in support to maize production during the period 2005–10 in the ten MAFAP countries grouped by subregion: Western Africa (Burkina Faso, Mali, Ghana, Nigeria), Eastern Africa (Ethiopia, Kenya, the United Republic of Tanzania, Uganda), and Southern Africa (Malawi, Mozambique). For the ten countries together, the policy stance was roughly neutral relative to international price levels between 2005 and 2007, with an average NRP of about zero. The average NRP to maize producers rose to 40 percent in 2008 before declining to around 20 percent in 2010.

However, this average trend masks differences in the trends among the various countries. Countries in Western Africa have provided higher levels of support to maize than those in Southern Africa, while countries in Eastern Africa have tended to tax maize production. Protection reached very high levels in 2008, particularly in Western Africa. This likely represents a policy response to the food price crisis, where governments put in place measures to support production, including through protective measures such as high tariffs and export bans in addition to productivity-enhancing measures such as input subsidies.

Figure 21 shows the percentage of maize production in the ten MAFAP countries that have received positive and negative protection respectively. Over the period 2005–10, the policies adopted have provided protection to an increasing share of maize production in the MAFAP countries (from 36 percent in 2005 to 66 percent in 2010).

The wide variation in levels of support and taxation for maize across the different countries and the sharp fluctuations from year to year suggest that considerable policy-induced distortion and uncertainty affects the incentives to invest in the sector.

FIGURE 19
Average relative rates of assistance by country, 2000-10*

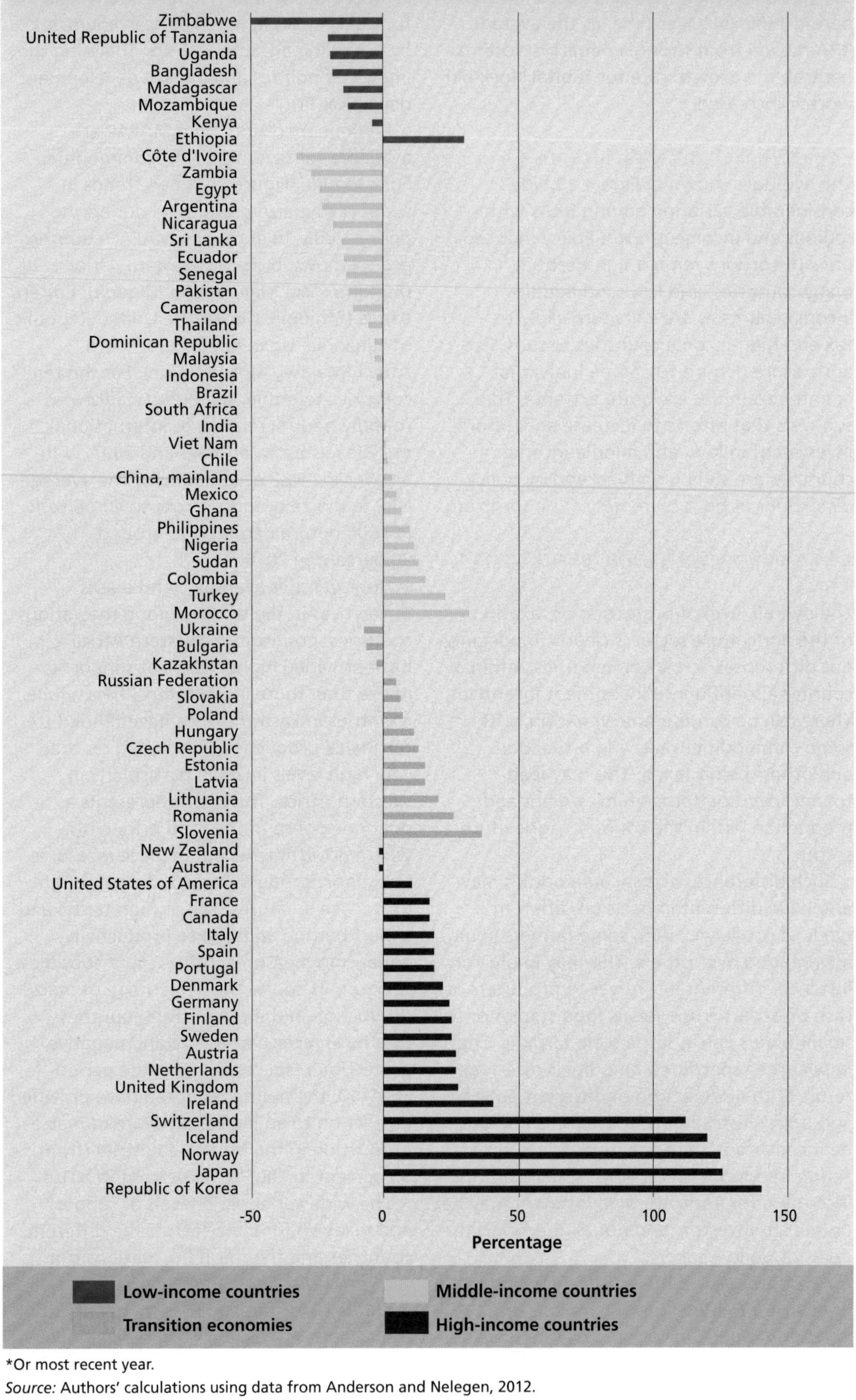

*Or most recent year.

Source: Authors' calculations using data from Anderson and Nelegen, 2012.

BOX 11
Monitoring African Food and Agricultural Policies (MAFAP)

MAFAP is working with national partners in ten countries in Africa to support decision-makers by systematically monitoring and analysing food and agricultural policies in the participating countries. The MAFAP analysis shows how domestic policy interventions – and sometimes excessive market access costs – affect incentives to farmers and their investment decisions. These measures are captured in estimates of nominal rates of protection (NRP),[1] which permit comparison over time and across countries. This analysis is supplemented by an analysis of the level, composition and effectiveness of public expenditures to determine the extent to which they are supportive of agricultural growth and development.

The information produced will feed into national decision-making processes and mechanisms for policy dialogue at the pan-African and regional levels, as well as to donors and other stakeholders. Efforts are made to embed the MAFAP activities in the country-level processes of the Comprehensive Africa Agriculture Development Programme (CAADP) so as to ensure that MAFAP results will be fully supportive of the overall CAADP endeavour towards agricultural development (see also Box 23 on page 87). MAFAP is also expected to become an element of the CAADP monitoring and evaluation framework and may provide useful benchmarks against which policy impacts can be analysed. The initiative is led by FAO in partnership with OECD and with major funding from the Bill and Melinda Gates Foundation.

[1] The NRP represents the increase or decrease in gross revenue from sales of a product relative to a situation of no policy intervention or excess market access costs. It excludes any possible increase in revenue resulting from direct budgetary transfers (such as input subsidies or taxes for example) and any other budgetary transfers not tied to production. Unlike the RRA discussed elsewhere in this report, the NRP does not consider the impact of policies protecting or taxing other sectors of the economy; it thus covers only that part of distortions to incentives that derive directly from policies affecting agricultural prices.

Potential gains from reducing policy distortions

This section has shown that policies in many countries at all levels of development influence the incentives to invest in agriculture, creating disincentives in many low- and middle-income countries and subsidies in many high-income countries. While these distortions have been reduced on average, they are still significant in many countries. Reducing the remaining price-distorting policies would improve incentives to invest and lead to better resource allocation by directing investment towards the activities and industries for which each country has its strongest comparative advantage. In the case of countries that still discriminate against agriculture, it is likely that such reforms would boost investment in agriculture, especially in the highest-payoff areas and subsectors. Several studies have estimated the impact of distorted incentives to agriculture on national and global economic welfare, economic growth and poverty.

Anderson, Valenzuela and van der Mensbrugghe (2009) provide a combined retrospective and prospective assessment in an economy-wide modelling exercise. They use the World Bank's global Linkage model (van der Mensbrugghe, 2005) to quantify the impacts of past reforms (up to 2004) and of potential benefits from removing remaining distortions in 2004. Their results confirm the significant gains to agriculture especially in developing countries from removing distortions to price incentives.

The dynamic effects of price distortions are analysed by Anderson and Brückner (2011), who examine econometrically the effect of moving the RRA towards zero on overall economic growth of sub-Saharan African countries. Given that most countries in the region currently tax agriculture, removing these price distortions would have

FIGURE 20
Nominal rates of protection of maize in selected countries of sub-Saharan Africa, by subregion

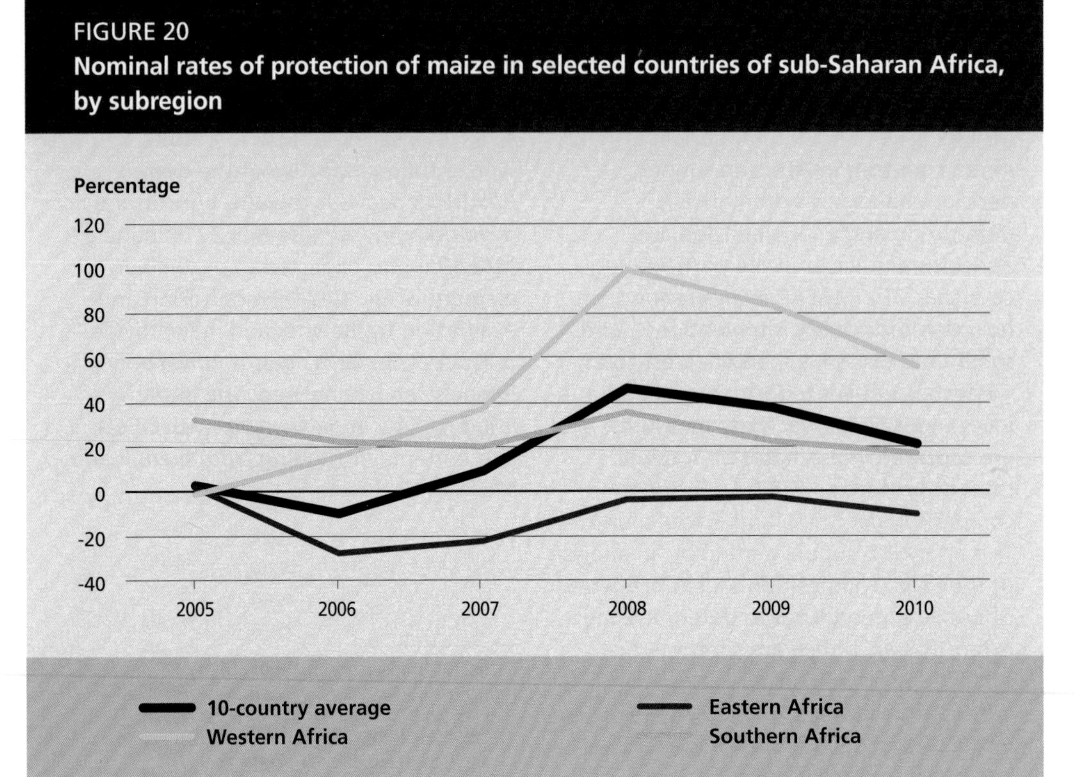

Notes: The ten countries included are focus countries in the MAFAP project. The averages shown are weighted by volume of production in individual countries.
Source: Short, Barreiro-Hurlé and Balié, 2012.

FIGURE 21
Share of maize production with positive and negative nominal rate of protection in selected countries of sub-Saharan Africa

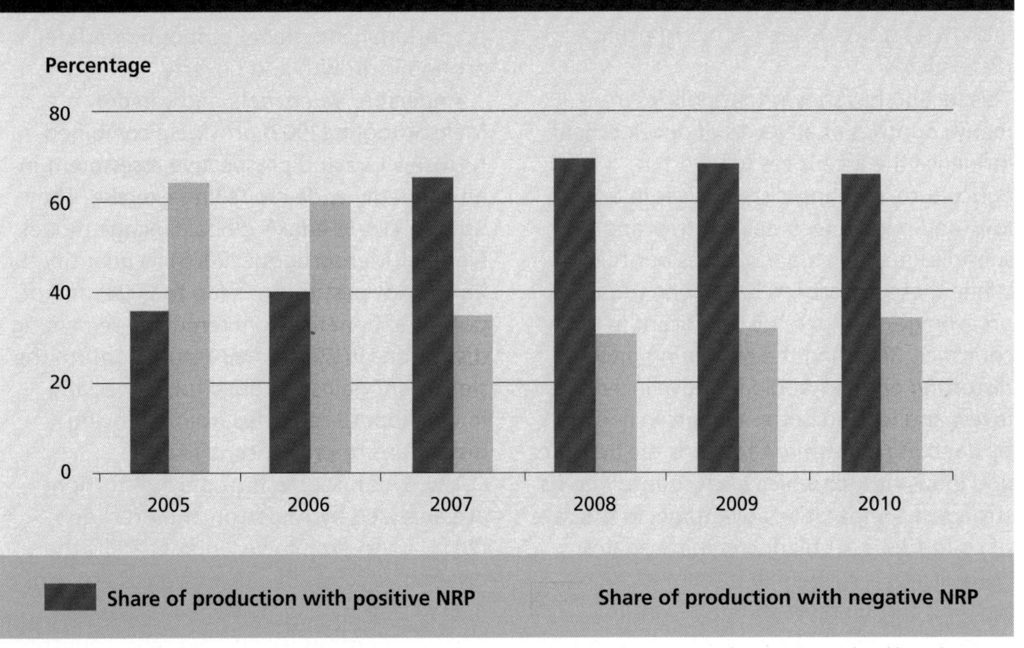

Notes: The ten countries included are focus countries in the MAFAP project. The averages shown are weighted by volume of production in individual countries. NRP = nominal rate of protection.
Source: Short, Barreiro-Hurlé and Balié, 2012.

a significant positive effect on their overall economic growth rate. These results show that taxing agriculture relative to other sectors reduces national economic welfare at a point in time and reduces overall output growth over time. Such model-based results are confirmed by experience in China (Box 12).

Including environmental costs and benefits in incentives to invest

Removing distortions is an essential element in improving the enabling environment for investment, but another element that will gain importance over time is the need to incorporate natural capital in investment decisions more effectively. The value of natural capital is normally excluded from assessments of agricultural capital (Box 13) and from national accounting, and both private and public investment decisions in agriculture have only intermittently and inconsistently accounted for the value of natural resources. Expansion in agricultural production over the last decades has been achieved at the cost of significant environmental degradation. The Millennium Ecosystem Assessment's (2005) comprehensive review of global environmental resources, for example,

BOX 12
Agricultural growth in China: the role of policies, institutions and public investment

In 1978, agriculture in China accounted for about 28 percent of GDP and 74 percent of the economically active population. The sector was mired in low rates of productivity, and rural poverty rates stood at 76 percent in 1980 (World Bank, 2007a). However, institutional reforms, market liberalization and investment – above all in research and development – kick-started a remarkable growth in agricultural productivity and rural incomes that ultimately led to industrialization and a massive reduction in poverty.

Reforms starting in 1978 focused on creating market institutions and incentives. Prices of agricultural commodities were raised, followed by institutional reforms that led to the farm household responsibility system (HRS). The reallocation of collective land for the use of households through contract arrangements with the collectives was a fundamental component of the HRS. Subsequently, markets were opened up in the mid-1980s (von Braun, Gulati and Fan, 2005). Reforms ultimately led to the steady dismantling of the state-run procurement and input supply systems and the creation of a largely market-driven system (World Bank, 2007a; von Braun, Gulati and Fan, 2005).

Institutional reforms, particularly the HRS and price reforms, are considered the dominant factors that led to increased productivity and annual growth in agricultural incomes of about 15 percent between 1978 and 1984 (McMillan, Whalley and Zhu, 1989; Lin, 1992; World Bank, 2007a; Fan, Zhang and Zhang, 2004). These were complemented by government investment, notably in agricultural research, development and extension, which tripled between 1984 and 2000 and strongly stimulated growth in agricultural production and poverty reduction (Fan, Zhang and Zhang, 2004) (see also Chapter 5).

Other types of investment also played an important role. For example, over the period 1953–78, China invested heavily in electricity. By 1998, about 98 percent of Chinese villages had access to electricity. This helped agricultural productivity growth and the establishment of township and village enterprises, which were a critical factor in the success of reforms (von Braun, Gulati and Fan, 2005). Investment in large-scale irrigation was also important: the proportion of cultivated area under irrigation increased from 18 percent in 1952 to about 50 percent in the early 1990s (Huang and Ma, 2010).

The experience of China shows the remarkable productivity and income gains that farming can generate when markets are allowed to operate, the incentives are right and public investments in technology and rural infrastructure are supportive.

BOX 13
Accounting for investment in natural capital

Natural resources are among the most important assets of developing countries. Investing in sustainable natural resource management is imperative for maintaining agricultural productivity, reducing risk of and vulnerability to natural disasters and ensuring provision of environmental services (such as hydrologic functioning, sediment control and biodiversity conservation). Yet natural capital is generally ignored in national accounts and in estimates of capital and are excluded from the United Nations System of National Accounts that is the basis for measuring GDP and other macro-level aggregates (except to the extent to which soil and water quality and water availability are capitalized into farmland values).

One approach to incorporating natural resource values and the costs and benefits of environmental services is "environmental accounting", which "provides a framework for organizing information on the status, use and value of natural resources and environmental assets ... as well as expenditures on environmental protection" (INTOSAI, 2010). However, few examples exist of attempts to include agriculture and land use in national-level environmental accounting systems, partly because of greater interest in the extractive sector in several countries, partly because of the sheer scale of agriculture and number of farms, and also because the complexity and heterogeneity of the ecosystems on which agriculture depends make truly comprehensive environmental accounting a daunting task. In March 2012, the UN Statistical Commission adopted the System of Environmental and Economic Accounting Central Framework as a recognized international statistical standard for environmental accounting.

concluded that 15 of the 24 global ecosystem services reviewed – including freshwater provision, climate regulation, air and water purification, natural hazard regulation and pest regulation – were being degraded or used unsustainably. In a world of gradually tightening natural resource constraints, ensuring the inclusion of environmental costs and benefits in the incentives faced by producers and investors in agriculture remains a key challenge.

Agriculture has multiple *impacts,* both positive and negative, on natural resources. In addition to producing food, fibre and fuel, the sector also produces a wide range of non-marketed outputs – externalities[17] – that result in broader costs and benefits to society that cannot be captured by farmers themselves (FAO, 2007; Morris, Williams and Audsley, 2007).

These outputs include some that result in net costs to society – greenhouse gas emissions, water pollution, soil erosion and degradation, groundwater depletion, etc. – as well as others that create benefits for society. These positive externalities include soil carbon sequestration, habitat creation and species protection, scenic beauty, flood control, recreational values and contributions to rural communities. In some cases, the value of the positive externalities created by agriculture can be enough to offset the costs (Buckwell, 2005).

Incorporating external costs and benefits into the incentives available to farmers and private investors and into the calculations underlying public investment decisions is crucial to ensure patterns of investment that are optimal from a social perspective. In deciding whether to clear a forest for use as cropland, most farmers would probably evaluate the costs of the labour and machinery required and the loss of any income derived from the forest against its projected value as cropland. Most farmers would lack incentives to consider the loss of forest carbon to the atmosphere, siltation of

[17] An externality refers to a situation where an individual's (for instance a farmer's) actions have unintended side-effects that benefit (positive externalities) or harm (negative externalities) another party. Both positive and negative externalities are pervasive in economic production, including in agriculture.

waterways downstream and loss of species habitat from the forest – and thus would disregard these factors.

In both public and private investment decisions, the problem is largely the same. Failure to consider the values to be assigned to natural resources and environmental goods and services – both costs and benefits – skews the investment decision by ignoring these goods and services. Addressing this problem involves major challenges. The sheer number of farmers in developing countries and the remoteness and poverty of many exacerbates the logistical difficulties and transaction costs involved. The complexity of agricultural ecosystems makes it difficult to accurately measure, quantify and monitor the biogeochemical and natural resource flows that underlie agriculture. There is a lack of analytical tools and mechanisms readily available to measure, value and account for resource use and loss in agricultural production systems. Several efforts are under way to remedy this situation, for example within the framework of the World Overview of Conservation Approaches and Technology (WOCAT) and the Land Degradation Assessment in Drylands (LADA) project.

Policy options for incorporating environmental values into investment decisions

A wide range of policy options are available for incorporating environmental values into investment and resource management decisions that are relevant for agriculture (FAO, 2007).

- **Command-and-control**. In this approach, governments use their regulatory powers to mandate certain behaviours, prescribe others and impose penalties for non-compliance. Command-and-control is the norm for pollution control in industrial settings, but the dispersed and fragmented nature of agricultural production makes such systems more difficult to implement.
- **Financial penalties and charges**. This approach modifies incentives through financial signals of taxes and fees. It does not prohibit certain activities but makes them more expensive to would-be polluters.
- **Removing perverse incentives**. In some cases, policy measures aimed at increasing agricultural production or productivity may unintentionally generate incentives to produce negative

BOX 14
Barriers to smallholder investment in sustainable land management

FAO recently conducted a review of the empirical evidence on the barriers to adoption of one important category of smallholder investment: sustainable land management (McCarthy, Lipper and Branca, 2011). Sustainable land management comprises agricultural practices such as agroforestry, soil and water conservation and grazing land management. A common feature of these practices is that they involve investment in ecosystem services to derive longer-term production as well as environmental benefits.

The review indicated that delayed benefits from these practices are a serious obstacle for many farmers. Up to five years may be needed to realize appreciable benefits, while costs are incurred immediately, partly in the form of opportunity costs arising from foregone income during initial phases of transition to sustainable systems. Lack of information and limited local experience with such techniques is a further deterring factor as it increases the uncertainty and risks involved in the investment.

On the other hand, well-functioning input supply and systems for managing collective resources such as pastures and waterways were found to have a positive impact on investment in sustainable land management. The review concluded that overcoming such barriers to widespread adoption of these techniques requires increased levels of public support, even though they generate higher returns to both farmers and the environment over the long run.

externalities. A classic example is subsidies on inputs, such as fertilizer or irrigation water, leading to excessive use, contamination of water through runoff and water depletion.
- **Establishing property rights to an externality**. This instrument relies on the privatization and allocation of rights to generate an externality, for example through permits to emit a defined quantity of air pollution or carbon. Such mechanisms often work in combination with other mechanisms such as payments for environmental services.
- **Payments for environmental services (PES)** encompass a wide range of instruments that involve various forms of payment for the provision of a positive environmental externality such as biodiversity conservation, watershed protection or climate change protection.

To the extent that environmental policies are applied to agriculture, command-and-control instruments or penalties and taxes have been the most common approaches. More recently, there has been increased interest in and development of payments for environmental services (PES). OECD (2010) notes the proliferation of PES programmes across developed and developing countries, mobilizing increasing amounts of finance and supporting international dialogues on efficient means of improving ecosystem services. The emergence of PES programmes is considered a promising approach that should be pursued by local and national governments as well as the international community (World Bank, 2007a). In the Global Environment Facility (GEF) and World Bank portfolios, PES schemes are increasingly being integrated into wider rural development and conservation projects, as a component for sourcing sustainable financing for investment (Wunder, Engel and Pagiola, 2008).

In spite of the interest in payments for environmental services, the number of functioning mechanisms in agriculture is limited. This is partly due to the numerous constraints – both conceptual and practical – still faced by such schemes (FAO, 2007; Lee, 2011). Policies and institutions that facilitate low transactions costs and the possibility for wide-scale replication are needed to realize the potential of this instrument to generate a significant and effective source of investment finance for sustainable agricultural development (Lipper and Neves, 2011).

Capturing opportunities to link to environmental finance

Resources available for investment in sustainable agricultural development can be augmented by linking to environmental public and private sources of finance (Lipper and Neves, 2011). The GEF is the largest public funder of projects to improve the global environment, providing grants for projects related to biodiversity, climate change, international waters, land degradation, the ozone layer and persistent organic pollutants.[18] An example from the private sector is the Livelihoods Fund, a mutual fund mobilizing 30–50 million euros from the private sector and foundations to finance programmes that contribute to both food security and carbon sequestration through the restoration of ecosystems. In return, investors receive carbon credits, which they can use as offsets or for sales. Linking climate finance to smallholder agricultural development is one of the objectives of the "climate-smart agriculture" approach (Box 15). However, the potential of environmental finance for smallholder agriculture development has been held back by the high transaction costs of measuring, reporting and verifying environmental benefits from small changes in a large number of agricultural operations; the lack of integration of such programmes into mainstream agricultural growth strategies; and the lack of legal and regulatory systems to create demand and willingness to pay for such services (Lipper and Neves, 2011).

Key messages

- Governments have the responsibility to support a favourable investment climate for agriculture, by creating an enabling environment and ensuring appropriate incentives for investment in agriculture. The well-known elements of an enabling environment for investment in general are equally relevant for agriculture: good

[18] Since 1991, the GEF has provided US$10.5 billion in grants and leveraged US$51 billion in cofinancing for over 2 700 projects in over 165 countries (GEF, 2012).

BOX 15

Linking climate and agricultural development finance to support sustainable agriculture development: the "climate-smart agriculture" approach

Climate-smart agriculture[1] seeks to support countries in increasing agricultural productivity and incomes, building resilience and the capacity of agricultural and food systems to adapt to climate change and reduce and remove greenhouse gases. Moving to sustainable and climate-smart agriculture will require higher levels of investment in human, social and natural capital. At the same time, changes in agricultural systems to increase sustainable growth can make a major contribution to sequestration, which could generate financial flows for the necessary investments (FAO, 2009a).

Achieving this requires actions at international and national levels. At the international level, climate financing commitments made in the Copenhagen Accord, for a goal of US$100 billion per year by 2020, must be fulfilled, and financing instruments that support the specificities of agricultural adaptation and mitigation must be established. At the national level, it is necessary to incorporate climate change adaptation and mitigation into national agricultural development strategies and investment plans. In Africa, incorporation of climate change issues into the CAADP provides an important platform for achieving this (FAO, 2012c). Also important is the building of national institutions, *inter alia*, to support the measurement, reporting and verification of adaptation and mitigation benefits from changes in agricultural systems that can serve as a basis for obtaining climate finance. (FAO, 2012c).

[1] Comprising crops, livestock, forestry and fisheries.

governance, the rule of law, political stability, low levels of corruption and the ease of doing business are strongly supportive of capital accumulation in agriculture. Governments that want to stimulate agricultural investment must get these basics right.

- Some elements are particularly important for agriculture, including secure property rights, rural infrastructure and public services, and market institutions. Vibrant agricultural input supply and agro-processing industries, which depend on an enabling environment, are also needed to ensure effective transmission of incentives to farmers.
- Government policies in agriculture and the broader economy can have a profound influence on the incentives – or disincentives – to invest in agriculture. Progress has been made internationally in reducing the policy distortions that have discouraged agricultural investment in many developing countries (relative to other sectors and other countries), but more needs to be done. Many low- and middle-income countries continue to tax agriculture heavily and, within countries, the unequal burden on different commodities may create additional uncertainty and disincentives for investors.
- Ensuring an appropriate incentive framework for investment also requires the inclusion of environmental costs and benefits into the economic incentives facing investors in agriculture. Many obstacles must be overcome in order to do this, including lack of analytical tools to measure and account for natural resources as well as the development of efficient institutions and mechanisms to lower transaction costs.

4. Promoting equitable and efficient private investment in agriculture

A favourable investment climate – consisting of an enabling environment for agriculture and appropriate economic incentives – is a necessary condition for stimulating and promoting more and better private investment in agriculture. However, a favourable investment climate is not sufficient to ensure that private decisions will achieve critical social goals such as greater equity and the eradication of poverty and hunger. Promoting socially equitable investment in agriculture requires additional measures to address the challenges faced by smallholders and to govern large-scale investment, thereby ensuring that the rights of local populations are protected and that they have the opportunity to benefit.

Low- and lower-middle-income countries typically have a large number of smallholder agricultural producers. These farmers are a crucial component of the agricultural economies of their countries. Some operate as commercial or semi-commercial enterprises, but many are subsistence or near-subsistence farmers who exist on the margins of survival. Smallholders can be more productive than larger farmers, but they often face particularly serious constraints that prevent them from effectively responding to better incentives for investment.

At the other end of the spectrum, large-scale corporate investors, including domestic and foreign corporations and sovereign investors, pose special challenges in low- and middle-income countries. Large-scale land acquisitions by foreign investors have received considerable recent attention, although large-scale domestic investors may be equally or more important. These large land acquisitions may represent a relatively minor share of total investment in agriculture or of total FDI, but they can have major impacts in the locations where they occur. Such investments may offer opportunities for employment, technology transfer and capital accumulation, but significant challenges exist in ensuring that they are respectful of the rights of local populations and offer real opportunities for smallholders to share in the benefits.

This chapter reviews some of the special issues involved in promoting and ensuring socially desirable outcomes of agricultural investment at these two ends of the spectrum of agricultural investors. It first discusses the importance of investments by smallholders and the specific constraints they face. This is followed by a discussion of the trend towards large-scale land-based investment and the issues involved.

Addressing the constraints to smallholder investors

Many factors justify a strong focus on better enabling smallholders to invest in agriculture, starting with their sheer numbers and economic importance and their relative productivity.[19] An estimated 85 percent of the 525 million farms worldwide are operated by smallholders on plots measuring less than 2 hectares (Nagayets, 2005). The evidence from a sample of six developing countries shows that more than 60 percent of rural people live on farms of less than the median size (Figure 22). In the same six countries, smallholder farms generate between 60 and 70 percent of total rural income through participation in farm and non-farm activities (Figure 23).

[19] While there is no unique and unambiguous definition of a smallholder, the most common approach is based on scale, measured either in absolute terms (2 hectares is standard) or relative to a country-specific threshold that takes into account agro-ecological, economic and technological factors. Definitions based on farm size ignore a number of other characteristics that are generally associated with smallholders, such as limited access to resources, reliance on family labour and less integration into markets.

FIGURE 22
Share of rural population by size of land holdings in selected low- and middle-income countries

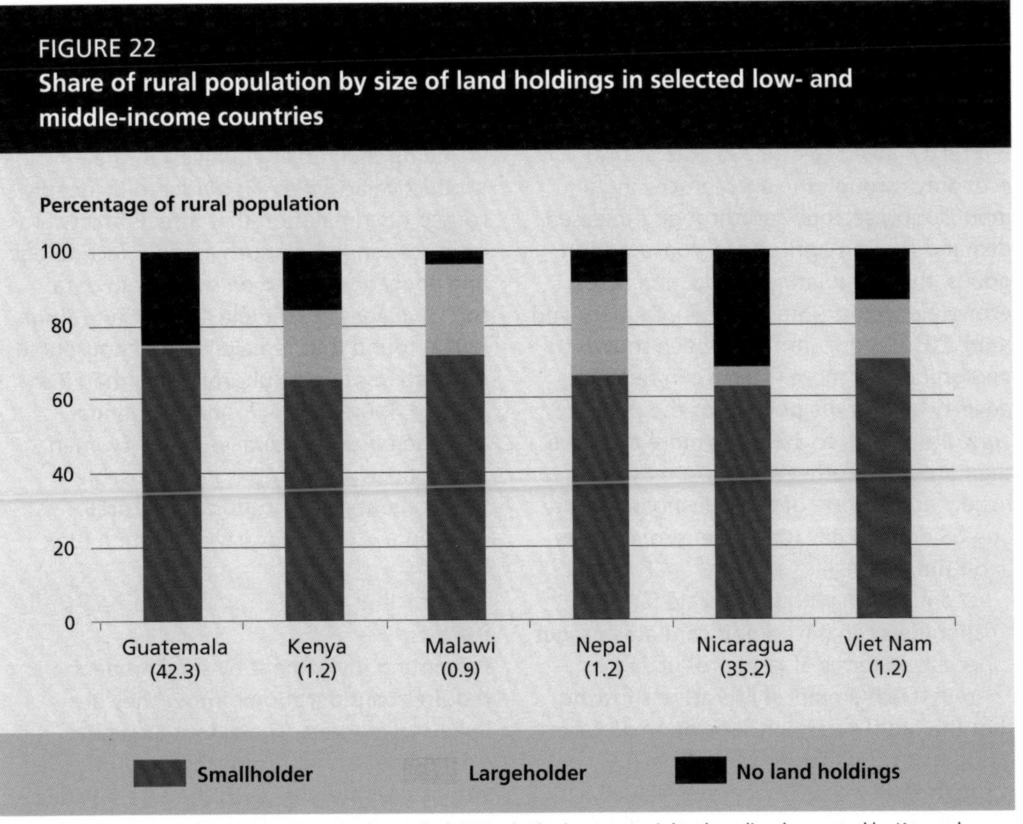

Notes: Farm size threshold indicated in parentheses (in hectares). The hectare-weighted median (suggested by Key and Roberts, 2007a and b) was employed as a threshold to classify smallholders and large farmers. The hectare-weighted median is calculated by ordering farms from smallest to largest and choosing the farm size at the middle hectare. Thus, half of all land (rather than half of all farms) is on farms smaller than the median.

Source: FAO, 2010b.

FIGURE 23
Share of rural income by size of land holdings in selected low- and middle-income countries

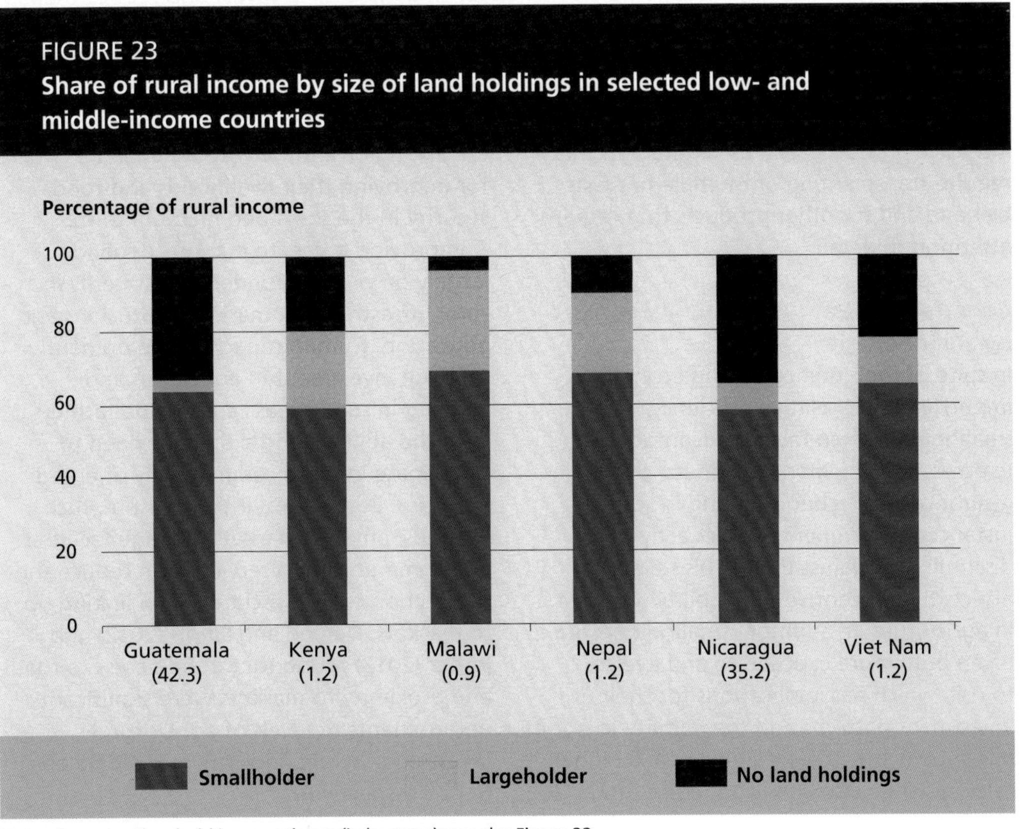

Notes: Farm size threshold in parentheses (in hectares); see also Figure 22.

Source: FAO, 2010b.

The contribution of smallholders to incomes in rural areas underlines their potential role as an engine of growth and poverty reduction. Smallholder income growth is directly linked to other sectors of the economy through the agro-processing and input supply sectors, and through increased demand for non-agricultural consumption goods, thus stimulating production across economic sectors (Christiaensen, Demery and Kuhl, 2010). Agriculture has been shown to be significantly more effective in reducing poverty among the poorest of the poor; agriculture is up to 3.2 times more effective than growth in other sectors in reducing headcount poverty of those living on below one US dollar a day (Christiaensen, Demery and Kuhl, 2010).

In comparison with large-scale farmers, smallholders can have significant advantages, especially in terms of land productivity. There is a rich empirical literature showing that the output per unit area on small farms is higher than on larger farms in many contexts (Eastwood, Lipton and Newell, 2010; Barrett, Bellemare and Hou, 2010). This results from greater intensity in the use of inputs, especially of family labour, and has positive consequences for food security. In general, the use of family labour when it is required offers flexibility denied to larger farms that depend on wage labour and it can reduce labour-supervision costs. Smallholder production is also more suitable for labour-intensive products, such as vegetables, that require transplanting or multiple harvests by hand and for other products that require attention to detail.

Constraints to agricultural investment by smallholders

In spite of their numerical and economic importance and relative efficiency, smallholders often face disadvantages in access to land, markets, inputs, credit, insurance and technology, and in some instances government policies actively discriminate against them. This severely affects their incentives and ability to invest in agriculture. In addition, smallholders are often both more exposed to and averse to risk, which has implications for their investment patterns and their ability to adopt investment strategies that may have higher returns, while also involving higher risk.

Many smallholders are women, for whom these constraints are, almost everywhere, even more severe (FAO, 2011d). Women's productivity and economic potential – including their ability to invest in their productive activities – is hindered by deeply rooted discrimination that affects access to resources and assets such as land, technology and education. Based on household data for 15 villages in Ethiopia, Dercon and Singh (2012) found that female-headed households invested less in agricultural assets than male-headed households. Closing the gender gap and ensuring equal access by women to resources and assets is indispensable for accelerating agricultural and rural development and poverty alleviation (Box 16).

Market linkages to facilitate smallholder investment

Promoting investment by smallholders requires consideration of how they are linked to markets. The extent to which smallholders produce for the market varies within the category, with the smallest farms producing primarily for home consumption and larger farms producing more for the market. Improving access to input and output markets can enhance the incentives for smallholders to invest and reduce their perception of risk. Increased investment, in turn, may boost their productivity and competitiveness. Even for farmers who produce primarily for home consumption, enhancing on-farm investment can be critical for improving their livelihoods and food security in the short and medium term.

Improving access to markets depends largely on publicly funded investments in rural infrastructure, market institutions and education. Human capacity development through investment in education and training in rural areas can provide farmers with the abilities and skills they need to participate in more commercially oriented activities. Better rural infrastructure, such as roads, physical markets, storage facilities and communication services, can reduce the transaction costs associated with linking up to markets. Dercon and Singh (2012) and Böber (2012) found that good access to roads and proximity to markets were significant determinants of levels of agricultural investment by smallholders in Ethiopia and Nepal, respectively.

BOX 16
Women are more constrained in agriculture

Women comprise on average 43 percent of the agricultural labour force in developing countries. The female share of the labour force ranges from about 20 percent in Latin America to almost 50 percent in Eastern and Southeastern Asia and sub-Saharan Africa. The share of rural household heads who are female, many of whom are farmers, ranges from about 15–40 percent in Latin America, 10–25 percent in Asia, and 20–45 percent in sub-Saharan Africa (FAO, 2011d).

Women farmers consistently have less access to the productive resources and services needed for farming than men: they are less likely than men to own land or livestock, adopt new technologies, use credit or other financial services or receive extension advice. For land, the most important asset for agricultural households, the available evidence shows that women represent fewer than 5 percent of all agricultural land holders in the countries of North Africa and West Asia for which data are available. In Southern Asia and Southeastern Asia, sub-Saharan Africa and Latin America the average is 12, 15 and 19 percent respectively.

Women are not only less likely to hold land; they also typically control smaller land holdings than men. Female-headed households have been found to own much less machinery than male-headed households. Livestock holdings of female farmers are also much smaller than those of men, and women are much less likely to own large animals, such as cattle and oxen, that are useful as draught animals. To this must be added significant differences in the education levels of female and male farmers, although access to education is one area where the gender gap has clearly narrowed in recent decades. The size of the gender asset gap differs by resource and location, but its underlying causes are repeated across regions: social norms systematically limit the options available to women.

Source: FAO, 2011d.

Governments also have an important role to play in addressing other key constraints to market participation. This can entail the delivery of important public goods and services that are not adequately provided by the private sector, such as research, development and extension, and market intelligence. Some of these can be provided by private agents, but will mostly require public funding. (See Chapter 5 for further discussion of public investment and expenditures).

Governments can play a more active role in leveraging private-sector participation in value chain development for export and domestic markets to the benefit of smallholders. Many mechanisms for alleviating the high transaction costs of market participation focus on the organization of smallholders into formal and informal groupings (see below for a discussion of the role of farmers' organizations).

Securing property rights and facilitating access to financial services

Insecure property rights, inadequate savings and limited access to financial services are critical constraints to smallholder investment. Insecure tenure for land, water and other resources can constitute a serious disincentive to invest in agriculture. This is particularly serious for women and other marginalized groups such as pastoralists and indigenous people. Secure property or tenure rights are necessary to provide incentives for longer-term investment, such as in land improvements. Clarity of tenure is necessary for landholders to make optimal investment decisions. Where rights are insecure, the balance of incentives to invest may be tilted away from agriculture towards other sectors. In addition to reducing the incentive to invest in agriculture, insecure property or tenure rights can also constrain access to financial services such as credit and insurance.

Limited access to financial services can severely constrain smallholder investment. Böber (2012), Dercon and Singh (2012) and Dias (2012) all found evidence of access to and/or cost of credit as a major factors conditioning on-farm investment by farm households (in Nepal, Ethiopia and Nicaragua, respectively, see Box 17). In many developing countries, the banking sector is oriented towards financing industry and trade, because the provision of financial services to regions with low population density and poor infrastructure is not profitable, due to high start-up costs, limited economies of scale and high transaction costs associated with the many small transactions typical of rural households when they save and borrow. The scarcity of financial services means that many rural households have very low saving rates and thus low levels of private investment.

In the past, governments have used subsidies to offset part of the fixed costs of providing rural financial services to producers; however, whereas specific one-time subsidies to financial institutions can be effective in overcoming the high start-up costs of financial operations in rural areas, generalized and continuing subsidies can be distorting and costly. Evidence suggests that subsidies to financial institutions in developing countries are often captured by middle-income households that already have access to banks, rather than benefiting

BOX 17
Empirical evidence on determinants of smallholder investment

There is a substantial theoretical and empirical literature on factors affecting smallholder productivity, but limited empirical evidence on how various factors specifically affect smallholders' investment decisions and their ability to invest. Three empirical case studies, were prepared for this report on Ethiopia (Dercon and Singh, 2012), Nepal (Böber, 2012) and Nicaragua (Dias, 2012). The studies looked at the relationship between farm investment and a range of other factors likely to affect investment. The conclusions emerging from the limited evidence of the case studies largely mirror findings relating to productivity and production.

One main conclusion confirms the local and context specificity of a range of factors affecting investment by farm households. A second important conclusion relates to the significance of community-specific factors determining overall investment by the community, while within communities a series of household characteristics determine how individual households respond to the overall local framework determined by the community characteristics. Among the community characteristics, the studies confirmed the importance of proximity to markets and access to transport infrastructure and credit. In terms of household characteristics, in general, wealthier and socially advantaged households were found to invest more than poorer and more disadvantaged households. Also, in some cases, male-headed households were seen to invest more than female-headed households. The studies suggest that providing infrastructure and promoting availability of credit are key contributors to promoting agricultural investment by relatively wealthier farmers. For poorer farmers in high-potential areas, however, this may be inadequate and further measures may be needed to help them escape the poverty traps that preclude them from expanding their assets.

In any case, the empirical evidence on determinants of farm household investment remains limited. More analysis is needed of constraints to smallholder investment and policy options to overcome them. In this context it should be noted that CFS at its 37th session in October 2011 requested the High Level Panel of Experts on food security and nutrition to undertake a comparative study of constraints to smallholder investment in agriculture in different contexts with policy options for addressing these constraints. Findings of the study are expected to be presented to CFS at its plenary session in October 2013.

BOX 18
Value chain financing for smallholders

There is growing interest in addressing finance through a value-chain finance approach. Agricultural value chain financing offers an opportunity to reduce cost and risk in financing and to reach out to smallholder farmers. Rather than assessing the potential borrower or investee, this approach takes a systemic viewpoint – looking at the collective set of actors, processes and markets of the chain. It is a transactions- and relationships-based assessment in which decisions about financing are based on the health of the entire system, including market demand, and not just on the individual borrower. A variety of potential financing mechanisms can then be applied according to the characteristics of the chain and its actors to ensure efficiency of financing, taking into consideration the costs, risks and investment capacity of the value chain actors. These in turn may pass financing up and down the value chain. In this way, many smallholders are able to secure funds that would otherwise not be available through conventional financing institutions, and agribusinesses are able to secure products and client loyalty that would likewise be difficult without the financing.

Source: Miller and Jones, 2010.

poorer rural households (Meyer, 2011; Claessens, 2005; Hoff and Stiglitz, 1997)

More effective approaches can be directed towards the development of value chains and the competitiveness of smallholders, allowing them more secure incomes and access to finance in kind or in cash through their value chain linkages (Box 18). Other instruments to be considered to enhance rural finance and investment may include support to new technologies to lower transaction costs of saving or borrowing, capacity building to both producers and financial service providers, tax breaks to financial institutions which provide services in rural areas and improvements in basic infrastructure.

Another aspect of financial services is related to risk insurance. Governments may intervene to assist in the provision of commodity price insurance because self-insurance strategies, such as crop and income diversification and consumption smoothing, may hinder investment and be inadequate to reduce income uncertainty. Market-based derivative instruments that provide insurance for internationally traded commodities are an important policy option (Larson, Anderson and Varangis, 2004). Market-based weather insurance that covers yield risks has also been suggested (Skees, 2008). Financial instruments such as futures prices and options allow producers to hedge against unforeseen price declines and reduce their exposure to income risk. In developing countries, risk management based on the use of such instruments will often require the involvement of marketing and financial intermediaries.

Building social capital to overcome constraints to investment

Smallholders need to build social capital if they are to take advantage of economic opportunities and incentives to invest and to overcome other constraints. Social capital can allow smallholders to engage more effectively in markets and with other economic actors and policy-makers, and can help compensate for lack of other assets such as land or financial capital. Effective and inclusive producer organizations can play an important role in this regard.

Rural producer organizations such as cooperatives can play a key role in strengthening the capacity of smallholders to invest in their agricultural activities. Depending on their mandate, their capacity and the specific context they operate in, they can take on different functions and forms, as well as provide a range of different services, thereby helping women and men producers overcome some of the critical constraints they face. They can also improve their incentives to invest and reduce and mitigate risk.

A broad variety of institutional arrangements have emerged in recent years. They provide smallholders with an array of services, ranging from enhancing management of natural resources, facilitating access to productive assets, markets and financial services and providing information and technologies to facilitating participation in policy-making.

Arrangements such as input shops (to collectively purchase inputs) and warehouse receipt systems (to collectively access credit) have increased smallholders' access to markets and productive assets, while reducing transaction costs. Mediation committees have improved smallholders' access to and management of natural resources. Producer organizations can be central in building small-scale producers' skills, providing them with appropriate information and knowledge and helping them to innovate and to adapt to changing markets.

Producer organizations can also help smallholders voice their concerns and interests and increase their negotiation power and influence on policy-making processes. Multi-stakeholder platforms and consultative forums are examples of mechanisms for smallholders to discuss the design and implementation of public policies.

Some key ingredients are needed for organizations to become effective and fully representative of the interests of smallholders. A recent collection of good practices (Herbel *et al.*, 2012) shows that successful organizations and institutional arrangements are the result of interdependent relations that smallholders develop and engage in:

- among themselves within the same organization (bonding);
- with similar organizations (bridging);
- through their organizations, with external actors (market actors, policy-makers, researchers, non-governmental organizations [NGOs]) within institutional arrangements (linking).

Through bonding relations, smallholders build close ties of solidarity at the grass-roots level. While bonding can be initiated by external support, evidence shows that such initiatives are more sustainable if initiated by the actors themselves. Bridging relations connect these groups together to form larger networks in the form of unions and federations of producer organizations and networks. Through bridging relations, smallholders enhance their access to assets and increase market and bargaining power.

To be fully effective, these organizations must also link with more powerful economic and policy actors, such as business corporations and the government. Relations with economic actors are important for smallholders not only to access markets but also to negotiate fairer commercial conditions. Collaboration with policy-makers is important to allow small- producers to participate in policy making and influence policy decisions.

In both developed and developing countries, there are examples of innovative producer organizations and institutional arrangements that have been successful in helping smallholders overcome different constraints. However, they too often remain limited in scale and scope. The main challenge is to build on these success stories in order to catalyse sustainable rural and agricultural development.

In order to scale up these successful initiatives, stakeholders must come together with clear roles and responsibilities to define the enabling environment for producer organizations to develop. The donor community and NGOs must focus on facilitating the development of existing producer organizations and cooperatives rather than on the introduction of new ones. Governments need to address the needs of existing smallholders and their organizations. Their support must be responsive rather than directive, investing in supporting these organizations to become effective.

In particular governments can provide the enabling conditions, which include policy, legal frameworks and economic incentives. Proactive measures are needed to promote the effective participation of women in mixed producer organizations and cooperatives by encouraging their leadership in these organizations. In addition, measures to support existing "women-only" producer organizations and cooperatives have proved to be a valuable strategy for women producers to develop their own producer organizations and cooperatives, based on their economic and social needs.[20] Consultative mechanisms for dialogue

[20] The Self Employed Women's Association in India provides an excellent example of a "women-only" organization that supports its members in achieving self-reliance through collective provision of a range of key services and building social capital (see FAO, 2011d).

between the government and producer organizations, allowing smallholders to fully participate in policy formulation, implementation and evaluation, are of utmost importance.

Social protection and smallholder investment in agriculture

Well-targeted social transfers can help many smallholders escape the poverty traps that prevent them from building assets. Social transfers are transfers of money designed to reach the most poor and vulnerable on a regular basis or in response to emergencies. For some poor households, transfers can represent a significant share of income and can help overcome or reduce the impact of two of the most serious constraints to investing and expanding household assets: lack of access to savings and credit and to insurance against risk (Barrientos, 2011). By providing liquidity, cash transfers can allow poor households to acquire different assets, including productive assets in agriculture (such as farm implements, land or livestock), as well as to invest in human capital through education. This can occur through an increase in the savings of poor households and/or by facilitating their access to credit. Programmes aimed at female household members can particularly help the acquisition of assets on the part of women, who tend to face even greater investment constraints than men.

Poor households in rural areas depend heavily on subsistence agriculture and have limited access to financial services such as credit and insurance. Social transfers to households can help them overcome this constraint and allow them to invest in productive assets. There is growing evidence of the positive impact of such programmes on growth and the productive and income-generating capacity of poor recipients (see Barrientos, 2011 for a review of some of the evidence). Social transfers can promote asset creation by households, protect against asset depletion in case of shocks and lead to improved investment decisions or resource allocation in general by providing some protection against risks (Hoddinott, 2008).

Evidence shows that participants in the Mexican *Oportunidades* social assistance programme invested 14 percent of their transfer payments during the first eight months – notably in farm animals, land for agricultural production and micro-enterprises, the latter mostly women-run – and after nine years beneficiary households had increased their consumption by 48 percent (Gertler, Martinez and Rubio-Codina, 2012). In Nicaragua, participants in the *Red de Protección Social* made fewer investments of this type, possibly because they were instructed to focus on food and education and possibly because of a lack of alternative economic opportunities in the region where the programme operated (Maluccio, 2010). Additional evidence regarding investment in productive assets by recipients of social transfers is found for the Bangladesh Rural Advancement Committee's *Challenging the Frontiers of Poverty Reduction – Targeting the Ultra Poor* programme (Ahmed *et al*., 2009; Barrientos, 2011). Also, Delgado and Cardoso (2000) found a high incidence of investment in productive capital among beneficiaries of the *Previdencia Social* Programme in Brazil.

Cash transfers can also help poor households tolerate risk and make more profitable investment decisions. Poor households often use productive assets as a buffer against shocks, which may lead them to prefer assets that can easily be converted to cash (Banerjee and Duflo, 2004). A high degree of risk aversion may also lead poor households to prefer types of investment with low risks but low returns over potentially more profitable but higher-risk activities. Cash transfers can give households more security and thus reduce their aversion to risk; they can also help households avoid detrimental strategies for dealing with shocks, such as selling productive assets or curtailing human capital formation by taking children out of school. In Nicaragua, where the *Red de Protección Social* operated during a severe economic downturn due to a record 30-year low worldwide price for coffee, Maluccio (2005) showed that programme beneficiaries were better able than non-beneficiaries to protect their income and human capital (by keeping children in school and maintaining access to basic health services). Sabates-Wheeler and Devereux (2010) report the same type of effects in Ethiopia as long as the shocks were not too severe relative to the size of the transfer.

Transfer programmes can have effects on the local economy beyond the immediate beneficiaries. Through the injection of a significant amount of cash into the local economy they can stimulate local product and labour markets through multiplier effects, thus also facilitating the creation of assets by non-participating households. Studies of rural pensions in South Africa (Møller and Ferreira, 2003) or in Brazil (Delgado and Cardoso, 2000; Schwarzer, 2000; Augusto and Ribeiro, 2006) strongly suggest such local-economy effects (Barrientos *et al.*, 2003). Similarly, at the community level, if transfers are provided through public works programmes, they can contribute to the creation of a series of productivity-enhancing public goods assets of importance for the local community.

A common question concerning transfer programmes is the potential reduction in household labour supply. Evidence from developing countries suggests that transfer programmes can reduce child labour, but there is little evidence to suggest that adult beneficiaries reduce their overall labour supply (Barrientos, 2011). From sub-Saharan Africa, Covarrubias, Davis and Winters (2012) and Boone *et al.* (2012) found that the Malawi transfer programme led to increased investment in agricultural assets, including crop implements and livestock, increased satisfaction of household consumption by own production, reduced agricultural wage labour and child work off-farm, and increased labour allocation to on-farm activities by both adults and children. For Ethiopian households with access to both the Productive Safety Net Programme (PSNP) and complementary packages of agricultural support, Gilligan, Hoddinott and Taffesse (2009) found no indication of disincentive effects on labour supply, but found that beneficiaries were more likely to be food-secure, borrow for productive purposes, use improved agricultural technologies and operate their own non-farm business activities. In a later study, Berhane *et al.* (2011) found that the PSNP has led to a notable improvement in food security status for those that had participated in the programme for five years versus those who only received one year of benefits.

Social transfer programmes thus seem to be a promising avenue to facilitate savings and investment by poor rural households, but there is a need for more research to understand more clearly the impact of transfer programmes – *inter alia,* on household asset accumulation and agricultural investment – and the implications for programme design.

Private cash transfers: the impact of remittances on farm investment

Emigration and remittances are significant phenomena in many countries. In Egypt, Morocco, Nigeria and Ethiopia, remittances account for between 5 and 10 percent of GDP (FAO, 2009b). The affinity of many migrants with agriculture often makes them more willing to invest in agriculture than in other areas. The emotional link of members of the diaspora with their communities of origin may imply a greater tolerance for investment risk. Furthermore, migration itself often results in lucrative export opportunities in the form of captured markets for "nostalgic goods" in diaspora communities. Migrants thus represent an innovative source of financing for agriculture specifically at the local level. Even when not invested directly in agriculture, remittances help mitigate risk, which facilitates adoption of new technologies and practices.

The exact impact of remittances on agriculture and smallholder farmers depends on the particular context. For example, in some rural areas of Morocco, emigration causes production to fall in the short term because of the withdrawal of labour from agriculture, while the long-term effects are positive as remittances are invested in agriculture (de Haas, 2007). Similar results were found in five Southern African countries; although domestic crop production falls initially, in the longer term crop productivity and cattle ownership are boosted by the inflow of remittances and higher domestic plantation wages (Lucas, 1987). In Ghana, the initial negative impacts of migration were completely compensated over time by remittances that stimulated both farm and non-farm production (Tsegai, 2004).

Evidence from Asia also shows positive longer-run effects of remittances. In the Philippines, Gonzalez-Velosa (2011) found that remittances were invested in working capital and also served as insurance. Farmers who received remittances were more likely to grow high-value crops, adopt hand tractors and threshers and invest in irrigation. There was no negative impact on production as there was no labour constraint to production. Overall, remittances have been found to facilitate agricultural development. In Bangladesh, Sen (2003) found evidence that off-farm labour, including migration, in combination with other diversification strategies, has allowed poor rural households to accumulate assets. Also for Bangladesh, Mendola (2008) shows that farmers with an international migrant in the family are more likely to adopt rice varieties with greater yield variability.

However, remittances do not always flow into productive investment in agriculture. In China, for example, de Brauw and Rozelle (2008) found that total grain output over 1986–99 fell about 2 percent as a result of migration, yet household disposable income rose by 16 percent. They reported that remittances were more often used for consumption rather than for productive investment. There is also substantial evidence that Mexican migrants are more likely to invest in housing than in productive activities (see references in de Brauw and Rozelle, 2008).

What determines whether remittances are invested in agriculture? A well-known study on Pakistan by Ballard (1987) concluded that unfavourable policies, such as central pricing, as well as poor infrastructure made investing remittances in agriculture unprofitable. Rather, remittances went into consumption and non-farm activities.[21] More recently, Miluka *et al.* (2007) found that Albanian households did not use remittances to invest in productivity-enhancing and time-saving farm technologies. As in Ballard's findings for Pakistani households, Albanian farm households expressed a desire to move out of agriculture, finding the policy context unsupportive.

Evidence from India supports the argument that agriculture attracts remittances for investment when farming is profitable. For example, Oberai and Singh (1983) found that in Punjab, a fertile area of India, remittances were invested in agriculture. On the other hand, evidence from Jharkhand, where only 30 percent of the land is cultivable, shows that only 13 percent of those owning 5–20 acres of land spent their additional income on farm productive uses (Dayal and Karan, 2003).

Making large-scale agricultural investment smallholder-sensitive

Trend towards large-scale land acquisitions

Large-scale private investment poses significant challenges for governments. Recent years have seen a surge of foreign acquisitions of land for agricultural use in developing countries. Land acquisition represents a transfer of ownership, but it does not necessarily add to agricultural capital in a country. Only if land acquisition is accompanied by additional capital assets such as land improvements, infrastructure, equipment or knowledge would it be considered investment from society's perspective. Thus, while land acquisition may offer opportunities for low- and middle-income countries to attract much-needed agricultural capital, the mere transfer of land is not sufficient. Such acquisitions can have serious implications for the affected communities, but the scale and impacts of such transactions are not always evident from media reports.

Data on land acquisitions based on in-country empirical research tend to show that the volume of officially recorded deals is well below that asserted in media reports, although the amount of land transferred can be large, and foreign entities typically constitute minority investors (Table 11). As an extreme case, more than half of all agricultural land in Liberia was involved in large-scale acquisitions between 2004 and 2009, but only about 30 percent involved foreign investors and much of it represented the continuation of long-standing concessions (Deininger and Byerlee, 2011). Significant shares of all agricultural land were involved in acquisitions in Cambodia (18 percent) and Ethiopia (10 percent), but domestic investors were responsible for the majority especially in recent years (Deininger and Byerlee, 2011; Horne, 2011). In most other countries the share of agricultural land involved in large-scale acquisitions was about 1–3 percent and foreigners were minority investors. Nevertheless, individual acquisitions can be very large. For example, Cotula *et al.* (2009) report that the maximum size of approved projects in the period 2004–09 in five African countries (Ethiopia,

[21] More recent research by Mansuri (2007) found that remittances are being invested in farm machinery, agricultural land, tractors and tube wells as well as human capital.

TABLE 11
Inventories of areas involved in large-scale land acquisitions

COVERAGE	LAND ACQUISITION	TOTAL AGRICULTURAL LAND, 2009	FOREIGN SHARE OF ACQUIRED LAND	TIME PERIOD
	(Million ha)		*(Percentage)*	
Country case studies				
Brazil[1]	4.3	265	..	Until 2008
Cambodia[2]	1.0	5.5	30	2004–09
Ethiopia[2]	1.2	35	51	2004–09
Ethiopia[3]	3.6	35	minority	2008–11
Liberia[2]	1.6	2.6	30	2004–09
Mali[4]	0.5	41	..	By end 2010
Mozambique[2]	2.7	49	47	2004–09
Nigeria[2]	0.8	75	3	2004–09
Sudan[2]	4.0	137	22	2004–09
Multiple countries				
Ethiopia, Ghana, Madagascar, Mali and Sudan[5]	2.5	270	..	2004–09
Mali, Lao People's Democratic Republic, Cambodia[6]	1.5	49	..	Until 2009
Kazakhstan, Ukraine, Russian Federation[7]	> 3.5	482	..	2006–11
25 countries in Africa[8]	51–63	800	..	Until April 2010
81 countries[9]	56.6	..	..	2008–2009
"Poor countries"[10]	15–20	..	..	2006–2009
Global studies				
Global[11]	15–20	4 900	..	Since 2000
Global[12]	70–200	4 900	..	2000–Nov. 2011

Notes: Studies use various methods for estimating the size of land acquisitions, including field visits, government documents, media reports and in-country research.
.. = data not available.
Sources: Hectares of agricultural land reported by FAO, 2012a. 1 FAO, 2011; 2 Deininger and Byerlee, 2011; 3 Horne, 2011; 4 Baxter, 2011; 5 Cotula *et al.*, 2009; 6 Görgen *et al.*, 2009; 7 Visser and Spoor, 2011; 8 Friis and Reenberg, 2010; 9 Deininger and Byerlee, 2011; 10 IFPRI, 2009; 11 von Braun and Meinzen-Dick, 2009; 12 Anseeuw *et al.*, 2012.

Ghana, Madagascar, Mali and Sudan) ranged from 100 000 ha in Mali to 425 000 ha in Madagascar.[22]

Recent land acquisitions have several distinctive characteristics, including (i) the involvement of international investors other than "traditional" multinational companies, (ii) their geographical origin, (iii) the large amount of land involved, (iv) the frequent lack of transparency and incompleteness of contracts and (v) the emergence of resource-seeking investors oriented to the production of food for export to their home markets (Cuffaro and Hallam, 2011).

In host countries, governments are generally engaged in negotiating investment deals (Deininger and Byerlee, 2011; Hallam, 2010). Agribusiness and industry account for the largest share of investors in land acquisitions, but foreign governments and sovereign wealth funds are increasingly involved in buying and/or leasing large tracts of farmland in the developing world.[23] Other investors expanding their exposure

[22] For an overview of land deals see the newly developed Land Matrix (http://landportal.info/landmatrix/index.php#pages-about).

[23] Sovereign wealth funds of China and the Republic of Korea along with the Gulf States of Qatar, Saudi Arabia and the United Arab Emirates appear to be emerging as key investors in these land purchases. Direct investment in foreign land at times occurs directly from government to government. In other occasions, the sovereign wealth funds work in conjunction with private sector intermediaries, their "private" subsidiaries or state-owned enterprises (McNellis, 2009).

in developing country agriculture include international private equity groups and international pension funds (McNellis, 2009; Anseeuw, Ducastel and Gabas, 2011; Davies, 2011; Wall Street Journal, 2010).

The drivers of large-scale land acquisitions seem to be different from those typical of foreign direct investment (Arezki, Deininger and Selod, 2011). The authors analysed the determinants of foreign land acquisition for large-scale agriculture from the perspective of both the country origin and the host country. From the side of the countries of origin, a main driving force is high dependence on food imports; from that of the host country, agro-ecological conditions are a main determining factor, with land acquisition more likely to occur in countries with ample supply of suitable land. In contrast with the general literature on FDI, the study finds a statistically insignificant relation between standard indicators of governance and land acquisition, indicating that overall levels of governance in the host country are not a strong determinant of such flows. Finally, and importantly, the authors find a significant *negative* correlation between an indicator of land governance and land acquisitions. Key variables included in the indicator are tenure security and recognition of existing land rights, the existence of a land policy and levels of land-related conflict. The implication is that weak land governance and poor protection of existing land rights in the host country may be a determinant of land acquisitions, either because investors favour countries with weak protection of land rights or because those are indeed the countries where such deals have been possible.

Currently, flows are unlikely to be large enough to have a marked impact at the global level. However, the impact – positive or negative – in some countries and localities can be considerable and warrants attention. A further factor calling for attention is the possibility of future growth in such flows; this, however, remains uncertain. At the same time, it should be noted that not all large-scale land acquisitions are financed from foreign sources. What is reported as a foreign acquisition is often partly domestic, frequently with more than half of the land acquired being owned by domestic investors.

The impact of large-scale agricultural investment

Land acquisition (and subsequent investment on the acquired land) represents one form of investment by large-scale corporate investors. Other forms may not involve direct control of land. The impact of such investment on recipient countries and affected local communities can be diverse, depending on the investment model chosen. On the one hand, large-scale corporate investment in agriculture can represent an opportunity. It can contribute to filling large investment gaps in poor countries with abundant natural resources but without the capacity to invest heavily in enhancing productivity. It can support the creation of infrastructure as well as the transfer of technology and know-how. Other potential benefits include the generation of employment and incomes as well as export earnings. However, investment involving acquisition of land can also be associated with major risks, including the possible neglect of rights of existing users of land, especially in the absence of strong governance and institutions for the protection of existing rights. Negative environmental impacts, *inter alia,* depletion of natural resources such as soil, water, forests and biodiversity, may also be significant threats.

Various recent initiatives aim to produce evidence on the implications for smallholders of large-scale agricultural investment. Within this context, an expert meeting on international investment in the agriculture sector of developing countries convened in November 2011 by FAO reviewed the current state of knowledge, including a series of case studies (see Box 19 for key results from one of these), on large-scale agricultural investment projects by both foreign and domestic investors (FAO, 2011f; FAO, 2012). These included different types of business models and different degrees of, and modalities for, involvement of local populations. Some involved acquisition of land by investors; others did not. The observed impacts were very diverse and depended on a range of factors.

Positive impacts at the national level included increases in agricultural production and yields, diversification of crops and, in some cases, higher export earnings as well as the adoption of higher standards where the investment targeted export markets.

BOX 19
Large-scale land acquisitions in Cambodia

Agriculture in Cambodia generates about 35 percent of the country's GDP (World Bank, 2012) and 65 percent of its employment (FAO, 2012a). Inflows of FDI have grown significantly, both overall and to agriculture. FDI to agriculture increased from an annual average of US$1 million in the period 2000–03 to an average of 53 million US$ in 2007–10.

Large economic land concessions (ELC), both foreign and domestic, have also been made, typically on 99-year lease contracts to enterprises for agricultural and/or agro-processing activities. Large tracts of land were leased already in the late 1990s and early 2000s (435 000 ha from 1999 to 2001), before the Land Law of 2001 and the Sub-decree on Economic Land Concessions in 2005 established a formal framework to regulate ELCs (including mandating environmental and social impact assessments and limits on the size of the land involved).

From 1995 to 2009, land involved in approved ELCs totalled about 1 million ha, a vast amount for a country with a total land area of about 17.5 million ha, 5.5 million of which are considered agricultural land (FAO, 2012a). The majority of conceded land involves domestically owned enterprises, with 35 percent going to foreign investors, mostly Chinese enterprises, followed in descending order by investors from Viet Nam, Thailand, the Republic of Korea and others.[1]

Preliminary impact analysis of seven agricultural projects active in 2010, each covering an area of agricultural land ranging from 4 000 to 10 000 ha, provided evidence of both benefits and costs. However, it is clearly not possible to ascertain to which extent the case study projects are representative of broader patterns in the country. All projects generated a large number of jobs and reported wages for unskilled workers far above the minimum wage for Cambodian garment workers. However, the benefits came at the expense of loss of land holdings and associated livelihoods by local communities. In some cases, there was evidence of environmental problems such as pollution or deforestation, although more in depth and comprehensive impact analysis would be necessary in order to draw firm conclusions.

One of the projects, a 4 000 ha rubber tree plantation, appeared to have been more successful in ensuring inclusiveness. It was characterized by a high degree of participation by the local community, continued ownership of much of the land by the local community and successful conflict resolution.

[1] The economic land concessions in Cambodia have come under severe criticism from civil society because of their impact on local populations and their environmental impact. According to a BBC report of 7 May 2012, the Government of Cambodia suspended the granting of land in order to curb eviction of local populations and illegal logging (BBC, 2012).

Source: Based on CDRI, 2011.

At the local level, one effect of FDI was the generation of employment. However, newly created jobs were often of limited duration and numbers. They were not always taken up by local people, and the net employment creation was limited when new jobs replaced former ones or self-employment. Some positive examples were also found of technology adoption and skills acquisition – in the case of outgrower schemes – as well as new or improved infrastructure.

Positive effects on the local economy were especially found in cases where the investment project was inclusive and actively involved local farmers, for example through outgrower schemes, contract farming or joint ventures. These include higher incomes for outgrowers selling products and services to the nucleus farm and the on-farm reinvestment of earnings by smallholders who have gained access to wage incomes.

On the other hand, the studies provided ample evidence of the possible negative impacts of large-scale land acquisition in countries where local land rights are not clearly defined and governance is weak. Negative social impacts included the displacement of local smallholders (often with inadequate or no compensation), the loss of grazing land for pastoralists, the loss of income for local communities and, in general, negative impacts on livelihoods due to reduced access to resources.

Also some evidence of negative environmental impacts was found, mainly higher pressure on natural resources due to intensification and reduction in forest cover and biodiversity. This was often due to the absence of proper prior environmental impact assessment and of effective environmental management systems during implementation. Nevertheless, some investment projects were found to have led to the adoption of environment-friendly technology.

Ultimately the studies indicate that the impacts on the local economy depend on a wide array of factors. Very importantly, they suggest that positive effects for local communities are unlikely to materialize when the investment involves land acquisition, especially when the land was previously utilized (including informally) by local communities. Other business models are much more likely to generate benefits for local populations.

Critical factors determining the impact – as opposed to the occurrence – are the policy, legal, and institutional framework in the host country and the capacity of host governments and local institutions to monitor and enforce contracts. At the local level, socio-economic conditions and the capacities of local civil society organizations, in particular farmers' organizations, are important. Impacts also depend crucially on the type of business model implemented, the terms and conditions of the contracts and the process of negotiation, design and planning of the investment project. From the side of the investor, important dimensions are the profile and priority objectives (e.g. speculation vs. long-term development) of the investor as well as the ability of local project managers to forge partnerships with the local community. A final key finding was the need for the presence of impartial and effective external support from third parties to ensure success.

Evidence also suggests that land-related agricultural investments have gender-differentiated implications (Box 20). Therefore, governments and international organizations that promote investment in agriculture need to specifically address gender – along with other social equity concerns – in policies and programmes relating to such investments.

Alternatives to land acquisition – more inclusive business models

Large-scale corporate investment in agriculture need not necessarily lead to the conversion of small-scale farming into large-scale agriculture. As suggested by the case-study evidence discussed above, other more inclusive partnership models exist that are more likely to achieve desirable developmental objectives by successfully combining assets of local farmers and investing companies. In such models, local farmers would provide land, labour and local knowledge, while corporate investors would provide capital, access to markets and technology and specialized knowledge. They would allow smallholders to make productivity-enhancing investment on their own farms.

A new trend offering opportunities in this respect is the fast-emerging development of investment funds for agriculture. Many of these focus their activities on agribusinesses and small and medium rural enterprises, with an emphasis on value addition through processing, logistics services, wholesaling, etc. Miller *et al.* (2010) analyse 31 investment funds and note the potential of such funds for increasing private sector interest in an area often considered too risky for many investors. Such funds can reduce the risk and difficulties faced by individual investors by pooling resources, diversifying across an array of agribusinesses and entrusting the portfolio to a professional fund manager. Many development agencies have also invested in these agricultural investment funds and commonly sponsor a parallel technical assistance facility to help ensure that investments can benefit small and medium enterprises and smallholders.

BOX 20
Gender implications of land-related investments in the United Republic of Tanzania

A case study of northern United Republic of Tanzania analysed gender-differentiated impacts and implications of corporate investments in jatropha production and horticulture.[1] The emphasis was on investments that were not based on large-scale land acquisition but adopted other business models involving farmers: group-based outgrower arrangements, individual informal and formal outgrower arrangements and permanent wage work.

The study found that the businesses examined were indeed creating new employment and income-generating opportunities for the rural population in the regions studied. It also found that there were gender-differentiated implications with respect to labour and income-generating opportunities for smallholders and wage workers. Some key findings were:

- Married women who were not outgrowers in their own right tended to see increased workload without benefiting equally from the investments, suggesting the need for income-generating opportunities targeted at women.
- The possibility for income generation of women outgrowers tended to be limited by their having fewer resources than men.
- On the other hand, women had equal and sometimes better access than men to formal wage employment in horticulture, but gender divisions of roles tended to lead to segregation between "men's" and "women's" work.
- Group-based outgrower arrangements in vegetable production offered both women and men better possibilities for income generation than casual labour on horticultural plantations and provided women in particular with a potentially expanding source of cash income to supplement existing income-generating activities and food production.
- Different types of crops may have different gender implications: indeed, women were found to have better opportunities than men for earning cash income from collection of jatropha seed, which has low profitability and is considered a "women's crop". Fewer women were able to access the more lucrative opportunities such as vegetable seeds, which requires more start-up capital.

Land-related investments were found to affect poor rural women and men differently in their access, use and control of land, *inter alia*:

- Contracting as outgrowers did not improve women's intra-household control and decision-making powers over use of land and income from it.
- Women contracting as outgrowers could generate supplementary income by renting-in additional land. However, this required availability of resources to start up the business.
- Women farming as wives of contracted outgrowers had enhanced decision-making power over land use, but for access and control they still depended on their husbands.
- Women involved in outgrower groups saw improved access to land and could avoid shifting land from own-food production to the outgrower crop.

The research also identified a series of specific good practices associated with each of the business models which can be incorporated into regulatory practices. The study pointed to the need to address constraints to women's access to outgrower activities and to the importance of special support to women outgrowers, including training and capacity building. It also concluded that group-based outgrower arrangements offered the significant benefit of self-employment, which participants in the study – especially women – valued above occasional employment opportunities on the horticultural plantations.

[1] The case study on the United Republic of Tanzania is the first in a series of case studies on the topic commissioned by FAO.

Source: Based on Daley and Park, 2011.

Vermeulen and Cotula (2010) provide a framework for analysing the nature of involvement of smallholders, operators and large investors in business models, consisting of the following four interlinked dimensions:

- Ownership: which stakeholders own the business and its key assets?
- Voice: who makes decisions in project design and execution?
- Risk: which groups bear the production, marketing and other risks?
- Reward: how are the costs and benefits distributed?

They describe six types of business models involving small-scale farmers in different ways (Box 21). In any case, there is no one perfect model, and there is also a large variety of situations, approaches and impacts within each business model. Whether a given business model benefits local development or not depends on many factors, including the local context.

The limited evidence on large-scale corporate investment reviewed above indicates that alternatives to land acquisitions, in which farmers keep or strengthen their control over land and which may create linkages to the surrounding economy, are more likely to provide benefits for all stakeholders. However, these benefits appeared to be neither automatic nor immediate. Many of the inclusive

BOX 21

Inclusive business models for corporate investment in agriculture

Alternatives to large-scale land acquisition, while not necessarily beneficial for all participants, include the following.

- **Contract farming** allows local farmers (or groups) to work their own land and enter a contract with a larger company to produce a given quality and quantity of agricultural produce by a certain date. The price is either agreed upon in advance or is based on a spot market. The company often provides up-front inputs to the farmers (seed, fertilizer, technical assistance, etc.).
- **Lease and management contracts** allow an agribusiness to lease land from small or medium scale landholders either for a fee or through a product or profit sharing agreement.
- **Tenant farming and sharecropping** arrangements involve small- or medium-scale farmers who lease land from large scale agribusinesses; in the former arrangement the farmer pays rent to the agribusiness and in the latter arrangement the farmer and the agribusiness agree on the fixed percentage of either profit or product which accrues to each party.
- **Joint ventures** include a very diverse set of arrangements whereby two or more stakeholders run the business. The partners share ownership, decision-making powers, risks and rewards, but they retain their individual legal status.
- **Farmers' organizations or cooperatives** are created by groups of farmers who form a jointly owned and democratically governed association to take advantage of economies of scale in business activities such as processing, storing or marketing products, as well as signing contracts and accessing finance. A response to the frequent criticism of slow decision-making is the incorporation of groups of farmers into – less democratic – farmer-owned companies, which – on the other hand – are able to make decisions more quickly.
- **Upstream and downstream business links** is a general term referring to arrangements that facilitate smallholders', operators' and agribusinesses' engagement in the manufacture of, procurement and or distribution of inputs to farming such as fertilizer, seeds, etc. (upstream activities) and/or the processing of agricultural products (downstream activities). Often they can facilitate international standards certification or other opportunities that are often not available to smallholders.

Source: Based on Vermeulen and Cotula, 2010.

investment models had faced various types of constraints and needed substantial initial external support (public and private). Such models may also imply higher transactions costs.

Experience with promoting win-win business arrangements in agricultural value chains shows the importance of intermediaries in bringing together smallholders and corporate investors. Intermediaries may be civil society organizations, specialized technical service providers, donors, but also government actors. According to the findings of the *Regoverning Markets* initiative, a facilitating and catalytic public sector is essential for the development of inclusive business models in modern agricultural markets, alongside a "receptive business sector" and organized farmers (Vorley and Proctor, 2008).

All stakeholders (governments, the international community, civil society and local communities) have an important role to play in helping to ensure the inclusiveness of agricultural business ventures. Governments, the international community and civil society can help address the power imbalances between local smallholders and large agribusinesses. Key actions in order to ensure socially and environmentally desirable outcomes for all stakeholders, especially smallholders, include (FAO, 2011e; Vermeulen and Cotula, 2010):

- Ensure that contracts are well developed, defined and enforced;
- Provide secure land tenure and fair compensation;
- Facilitate the recognition of land as equity for credit;
- Improve access to banks, insurers, law firms and courts;
- Educate and raise awareness regarding business operations and access to market information;
- Facilitate a participatory process that empowers smallholders and locals;
- Empower locals to form farmers' organizations;
- Increase transparency and information (including documentation) regarding FDI and land acquisition;
- Encourage *ex-ante* and *ex-post* monitoring and evaluation of social, gender and environmental impacts.

Governance to improve the social and environmental impact of investment in agriculture

With a view to providing guidance on how to ensure more desirable agricultural investment, together with other stakeholders (including the international community, governments, private sector, civil society and academia), FAO has pursued mutually supporting frameworks such as *The Voluntary Guidelines on the Responsible Governance of Tenure of Land, Fisheries and Forests in the Context of National Food Security* (VGGT) and *Principles for Responsible Agricultural Investment that Respects Rights, Livelihoods and Resources* (PRAI).

The VGGT are intended to serve as a reference by setting out principles and internationally accepted standards for responsible practices for tenure and its governance (FAO, 2012b). They provide guidance on a wide range of areas, including the development and implementation of policies and laws, the administration of tenure, and environmental issues such as climate change and natural disasters.

The VGGT set out ways in which governments and other stakeholders might best ensure that FDI and other investment have socially and environmentally desirable impacts. They encourage responsible investment where tenure is affected, with a view to improving food security. They identify safeguards that should be in place so that investment, particularly deals involving large-scale acquisition of land, recognize and protect the existing tenure rights of potentially affected people and communities. They provide guidance on areas such as ensuring a consultative and participatory process of negotiations among investors and other stakeholders.

The VGGT are based on an inclusive process of consultation, where government officials, and representatives of civil society, private sector, research organizations, UN bodies with a mandate in the field of food security and nutrition and academia identified and assessed issues and actions. The Guidelines were finalized through inclusive consultations and intergovernmental negotiations led by CFS and officially endorsed by a Special Session of CFS on 11 May 2012.

In addition, FAO, IFAD, UNCTAD and the World Bank have also formulated seven key principles for what constitutes *Principles for Responsible Agricultural Investment that Respects Rights, Livelihoods and Resources* (PRAI) (FAO, 2011g, FAO *et al.*, 2012). The overriding objective is to ensure that investment in agro-enterprises results in a mutually beneficial outcome for all. They offer a framework that can be used as a basis for formulating laws, regulations, investment contracts, international agreements or corporate codes of conduct but do not define a specific monitoring system. However, some civil society groups have publicly criticized the RAI principles as being too weak (FIAN, 2010 and Transnational Institute, 2011), in particular due to their limited link to human rights.

The broad principles for responsible agricultural investment formulated by the four agencies are:

- **Land and resource rights.** Existing rights to land and natural resources are recognized and respected.
- **Food security.** Investments do not jeopardize food security but rather strengthen it.
- **Transparency, good governance and enabling environment.** Processes for accessing land and making associated investments are transparent and monitored and ensure accountability.
- **Consultation and participation.** Those materially affected are consulted, and agreements from consultations are recorded and enforced.
- **Economic viability and responsible agro-enterprise investing.** Projects are viable in every sense, respect the rule of law, reflect industry best practice and result in durable shared value.
- **Social sustainability.** Investments generate desirable social and distributional impacts and do not increase vulnerability.
- **Environmental sustainability.** Environmental impacts are quantified and measures taken to encourage sustainable resource use, while minimizing and mitigating their negative impact.

The CFS Bureau and its Advisory Group supported by the joint Secretariat has begun an inclusive multi-stakeholder consultation process for the development and broader ownership of principles for responsible agricultural investment that enhance food security and nutrition. The consultation process will ensure consistency and complementarity with the VGGT. The PRAI and related research outputs will be considered as inputs to this process.

Key messages

- A favourable climate to foster private investment in agriculture is indispensable for all investors, but is not sufficient to allow all farmers to invest in their productive activities and to ensure that private investment meets socially desirable goals.
- Smallholders require special attention in order to allow them to overcome the constraints they often face to invest, including poor access to markets and financial services, insecure property rights and vulnerability to risk. Supporting the formation of social capital in the form of effective producers' organizations and providing social transfer programmes allowing them to build assets can help overcome some of the constraints.
- Large-scale investment in agriculture may present opportunities but land acquisition also poses special challenges in terms of potential impacts on smallholders and the rural poor. It is important to improve the governance of large-scale investment and promote inclusive business models that allow local populations to benefit.
- Both cases underscore the indispensable role of government in ensuring an appropriate enabling environment for socially desirable private investment and in investing in essential public goods.

5. Channelling public investment towards higher returns

Public investment in agriculture is required to foster more and better private investment and to ensure that it is economically and socially beneficial. Public goods for agriculture, such as R&D, education and rural infrastructure, are a fundamental part of the enabling environment described in earlier chapters and they are essential for agricultural growth and poverty reduction. However, governments everywhere face financial constraints and competing demands, so they must make difficult choices in allocating public resources. Which public investments have the highest returns in terms of agricultural growth and poverty reduction?

Evidence shows that investment in public goods have much higher returns than other expenditures such as general subsidies, but what constitutes a public good is not always clear-cut and may differ by context. Even though some types of investment are known to yield high economic and social returns, they are not always given the highest priority in budget allocations. Understanding the impact of different types of public investment and expenditure on agricultural performance and poverty alleviation can help guide public investment towards higher returns.

Returns on public investment in and for agriculture[24]

Early studies of the impact of aggregate agricultural expenditures on growth and poverty reduction found diverging results. One of the earliest studies in this field (Diakosavvas, 1990) found that government expenditure on agriculture had a strongly positive effect on sector performance, but a comparative analysis of data for 100 countries failed to find a statistically significant effect of agricultural spending on growth in per capita GDP (Easterly and Rebelo, 1993).

More recent studies have highlighted that the type of expenditure matters. Public expenditures on agriculture, education and roads contribute strongly to agricultural growth across regions, although to different degrees; and, within agriculture, the impact of research expenditures on productivity is stronger than non-research expenditures (Fan and Saurkar, 2006). Investment in research, often associated with extension, is consistently found to be the most important source of productivity growth in agriculture (Fischer, Byerlee and Edmeades, 2009).

Also country studies in several regions have found positive relationships between government expenditure on agriculture and growth in agricultural and total GDP, while confirming that the type of expenditure matters. In Rwanda, for example, 1 dollar of additional government expenditures on agricultural research increases agricultural GDP by 3 dollars, but the effects were larger for staples such as maize, cassava, pulses and poultry than for export crops (Diao *et al.*, 2010). In India, expenditures aimed at improving productivity in livestock had greater returns and were more effective in mitigating poverty than general public investment in agriculture (Dastagiri, 2010).

The substantial literature on public investment in agricultural research and development (R&D) shows that it has been one of the most effective forms of public investment over the past 40 years. Because R&D drives technical change and productivity growth in agriculture, it raises farm incomes and reduces prices for consumers. The benefits multiply throughout the economy as the extra income is used to purchase other goods and services, which in turn create incomes for their providers. The welfare effects are large and diffuse, benefiting many people who are far removed from agriculture, so they are not always recognized as stemming directly from agricultural research (Alston *et al.*, 2000; Fan, Hazell and Thorat, 2000; Evenson, 2001; Hazell and Haddad, 2001; Fan and Rao, 2003).

[24] This section is based on a background paper prepared by IFPRI staff members. See Mogues, *et al.*, 2012.

In a review of 375 applied research programmes and 81 extension programmes, Evenson (2001) found that in four-fifths of the applied research programmes and three-quarters of the extension programmes the reported rates of return were greater than 20 per cent and that in many they exceeded 40 percent. Alston *et al.,* (2000) reviewed 292 studies covering 1953 to 1997 and found average rates of return on agricultural research of 60 percent in developing countries. In an update of that study, Alston (2010) found the global rate of return to R&D to have been consistently high.

Recent country level studies support the findings of these comprehensive reviews. For example, research in Thailand is estimated to have a significant positive impact on TFP and a marginal rate of return of 30 per cent (Suphannachart and Warr, 2011). Analysis of an extension service in Uganda reveals rates of return of between 8 and 36 per cent (Benin *et al.,* 2011).

Ensuring enhanced expenditures on agricultural R&D is clearly a priority. As discussed in Chapter 2, higher-income countries have significant private R&D expenditures, but in developing countries most R&D efforts are publicly funded. Public-private partnerships constitute an innovative approach to involving the private sector both in R&D efforts as well as in the provision of other public goods (Box 22).

Returns to complementary investments for agriculture

Investment in rural public goods such as education, rural infrastructure, health and social protection measures can generate important benefits for the agriculture sector and for its contribution to economic growth and poverty alleviation. Rural public goods are complementary; investing in one often enhances investment in the other. Evidence also shows that agricultural productivity and poverty reduction are compatible goals; investing in rural public goods tends to have high payoffs for both.

Studies have compared the impact on both agricultural performance and poverty of public spending on agriculture with that of other forms of expenditure. Figures 24 and 25 summarize results for such analysis undertaken in four developing countries: China (Fan, Zhang and Zhang, 2004), India (Fan, Hazell and Thorat, 2000), Thailand (Fan, Yu and Jitsuchon, 2008) and Uganda (Fan and Zhang, 2008). The impact of public investment on the value of agricultural production was consistently the highest for agricultural research and development.

After agricultural R&D, the ranking of returns to other investment areas differs by country, suggesting that public investment priorities depend on local conditions, but rural infrastructure and road development are often ranked among the top sources of overall economic growth in rural areas (Fan, Hazell and Thorat, 2000; Fan, Zhang and Zhang, 2004; Mogues, 2011). In Ethiopia, access to all-weather roads reduced poverty by 6.9 percent and increased consumption growth by 16.3 percent (Dercon *et al.,* 2009). Mogues (2011) found that returns to public investment in road infrastructure in Ethiopia were by far the highest of all categories. In Uganda, the marginal returns to public spending on feeder roads on agriculture output and poverty reduction was three to four times larger than the returns to public spending on larger roads Fan and Zhang (2008).

Public goods in rural areas also tend to be complementary. For example, in Bangladesh, villages with better infrastructure benefited more from agricultural research than villages with poorer infrastructure; they used more irrigation, improved seed and fertilizer, paid lower fertilizer prices, earned higher wages and had significantly higher production increases (Ahmed and Hossain, 1990). In Viet Nam, rural roads fostered the development of local markets and raised primary school completion rates, improving incentives for agricultural investment as well as investment in human capital (Mu and van de Walle, 2007).

In a classic assessment of international cross-country evidence, Antle (1983) found that lack of transportation and communication infrastructure posed severe constraints to aggregate agricultural productivity in developing countries, suggesting that investment in these areas would have high payoffs for agriculture. This conclusion was supported by results from India (Binswanger, Khandker and Rosenzweig, 1993), Colombia and Thailand (Kessides, 1993; Binswanger, 1983), and Nepal (Jacoby, 2000). Other more recent studies have confirmed the positive impact of investment in transport and communications infrastructure on agricultural growth; a cross-country comparison found that a 1 percent increase in government spending

BOX 22
Public-private partnerships

Public-private partnerships have received increasing attention as a way to involve the private sector in supplying goods and services with some degree of public goods characteristics and for bringing together private and public investors to promote agricultural development, poverty reduction and food security. For example, in May 2012 the Grow Africa Investment Forum emphasized the need for the formation of new agricultural partnerships between the public sector, private sector and communities.

Public-private partnerships are generally defined as the participation by the private sector in an economic activity in which the parties involved share costs, risks and benefits but where, if left to the free market alone, such private activity would not occur due to low private returns to investment or the high level of risk involved (Warner, Kahan and Lehel, 2008). Several examples of such partnerships are found in farm-to-market roads, water for irrigation, wholesale markets and trading centres, agro-processing facilities and information and communications technology. Each type of public-private partnership offers specific benefits and challenges. Public-private partnerships specifically for sustainable agricultural development can also include a variant known as "hybrid value chains" (Drayton and Budinich, 2010; Ferroni and Castle, 2011), which are multi-partner structures that bring together private companies with entities such as non-governmental organizations, university research institutes and foundations. Another type of public-private partnerships involves collaboration among public and private entities for undertaking research, developing new technologies and creating new products to benefit resource-poor farmers and marginalized groups in developing countries (Spielman, Hartwich and von Grebmer, 2007).

Effective strategies for use in agricultural development

Many new examples of public-private partnerships have developed over the last several years; such partnerships have been forged to undertake projects in areas such as agricultural productivity, biofortification, technical and investment assistance and export strategy. Major examples include

in these areas raised agricultural GDP growth by 0.01 to 0.14 percent (Benin, Nin-Pratt and Randriamamonjy, 2007).

Different types of public investment in rural areas also have strong poverty-reducing impacts (Figure 25). The rankings are slightly different from those for agricultural productivity but agricultural R&D, education, roads and electrification rank highly in all countries for both goals. The implication is that agricultural growth and poverty-reduction objectives are strongly compatible objectives; investment in rural public goods tends to have positive impacts on both.

Returns over time to investment in agriculture

Returns to many types of agricultural spending have declined over time, but returns to investment in agricultural R&D have remained high. In India, returns to expenditures on agricultural credit subsidies were fairly high in the 1960s and 1970s, but they declined sharply over time while the returns to agricultural R&D have remained high for decades (Figure 26) (Fan, Gulati and Thorat, 2008). A comprehensive review of R&D and extension found that internal rates of return were as high in the 1990s as they were in the 1960s (Evenson, 2001). Likewise, for Thai crop production, public spending on research was a positive and significant determinant of TFP growth from 1970–2006 (Suphannachart and Warr, 2011).

In the long run, returns to spending on agricultural subsidies have fallen behind those for R&D, roads, education and irrigation infrastructure. In India, the overall poverty-reducing impact of agricultural expenditures has declined as a result of the declining share of agriculture in the economy and the increase of the proportion of agricultural expenditure given in the form of subsidies (Jha, 2007).

the Southern Agriculture Growth Corridor of the United Republic of Tanzania, the HarvestPlus Challenge Program as well as those currently facilitated by the creation of such organizations as the Ghana Commercial Agriculture Project and, in Nepal, the Agro Enterprise Centre.

Common elements of success attributed to these public-private partnerships generally include project plans with clearly-defined objectives, roles and responsibilities, milestones, risk management and mitigation strategies, as well as the provision of in-kind rather than cash only contributions from private sector partners. Effective and efficient definition of and implementation of local government policies is also crucial (Spielman, Hartwich and von Grebmer, 2007). The HarvestPlus Challenge Program is trying to implement these success factors in its current multi-partnership effort.[1]

Challenges

Spielman, Hartwich and von Grebmer (2007) present the results of a study[2] examining how public-private partnerships in agricultural research stimulate greater investment in pro-poor innovation in developing country agriculture. Challenges arise through the creation of hidden transaction costs despite the overcoming of the prohibitive costs of conducting research or deploying products independently. Although not easily quantifiable (see also Warner, Kahan and Lehel, 2008) these can pose significant barriers to success. In addition, it was shown that few of these partnership projects have the adequate risk management or mitigation strategies in place. Other concerns such as internal conflict resolution and legal and financial strategies, if not clearly defined, also tend to threaten the value produced in these public-private partnerships.

[1] This is a multi-partner collaboration in biofortification supported by the Syngenta Foundation for Sustainable Agriculture, which focuses on improving the nutritional value of staple foods. Although still in the development phase until 2013, it has released one crop already available in Uganda and Mozambique (Ferroni and Castle, 2011).

[2] The study examines 75 projects undertaken by the Consultative Group on International Agricultural Research (CGIAR) in partnership with various types of private firms operating on national, regional and international levels.

The effects of public expenditures on poverty reduction also tend to decline over time (Figure 26). Furthermore, the poverty-reduction impact of public subsidies for fertilizer, irrigation, power and credit are consistently well below that of public spending on R&D, education and roads. Although subsidy expenditures are frequently rationalized based on equity and poverty considerations, these results indicate that investment in public goods is clearly more effective in this regard.

Returns to investment in more-favoured versus less-favoured regions

Returns to public investment in agriculture are likely to differ according to location. A long-standing policy debate concerns whether it is better to invest public resources in more-favoured areas with higher agro-ecological potential, or in less-favoured areas, where poor populations tend to be concentrated.

Regions that are well-endowed with favourable agro-ecological conditions and easy access to markets seem like the obvious place when aiming to raise agricultural productivity. Investing in these more-favoured, high-potential regions may also be an effective strategy for reducing poverty because it offers "spillover" and "multiplier" benefits to residents of more remote regions who may migrate to take advantage of employment and income opportunities in the more-favoured region (Palmer-Jones and Sen, 2003).

On the other hand, targeting less favourably endowed agro-ecological regions may yield higher returns, at least in terms of poverty reduction, because the marginal costs of achieving further gains in well-endowed regions increase over time after the easy gains have been achieved (Ruben and Pender, 2004). Only few empirical studies have addressed the issue of returns to investment in more-

FIGURE 24
Returns to public spending in terms of agricultural performance

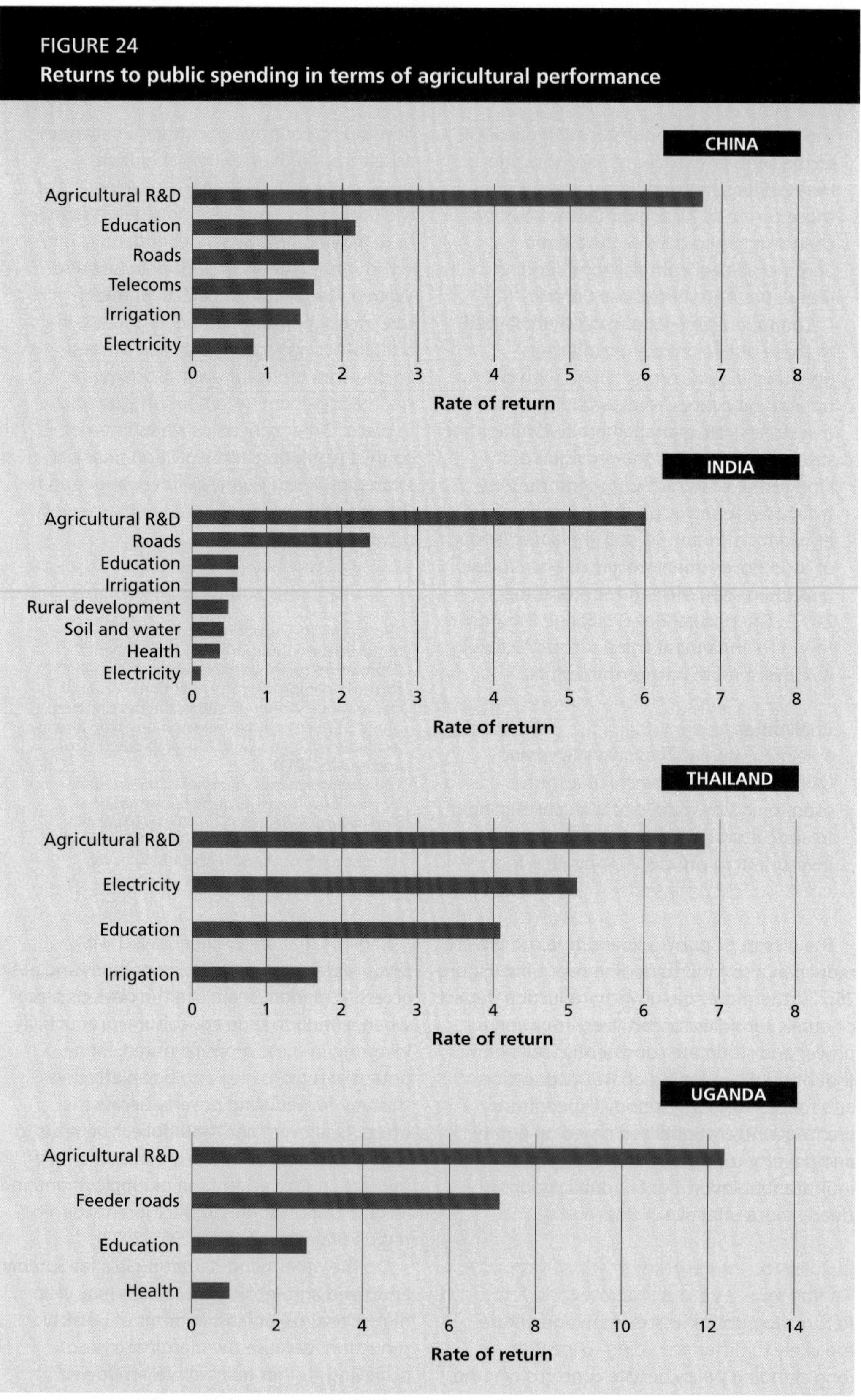

Notes: The magnitudes are returns to one monetary unit of different types of public spending in terms of increased agricultural production or productivity measured in the same monetary unit. The agricultural performance variable is measured slightly differently in each country: agricultural GDP in China, agricultural total factor productivity in India, and agricultural labour productivity in Thailand and Uganda.
Sources: Fan, Zhang and Zhang, 2004; Fan, Hazell and Thorat, 2000; Fan, Yu and Jitsuchon, 2008; Fan and Zhang, 2008.

FIGURE 25
Returns to public spending in terms of poverty reduction

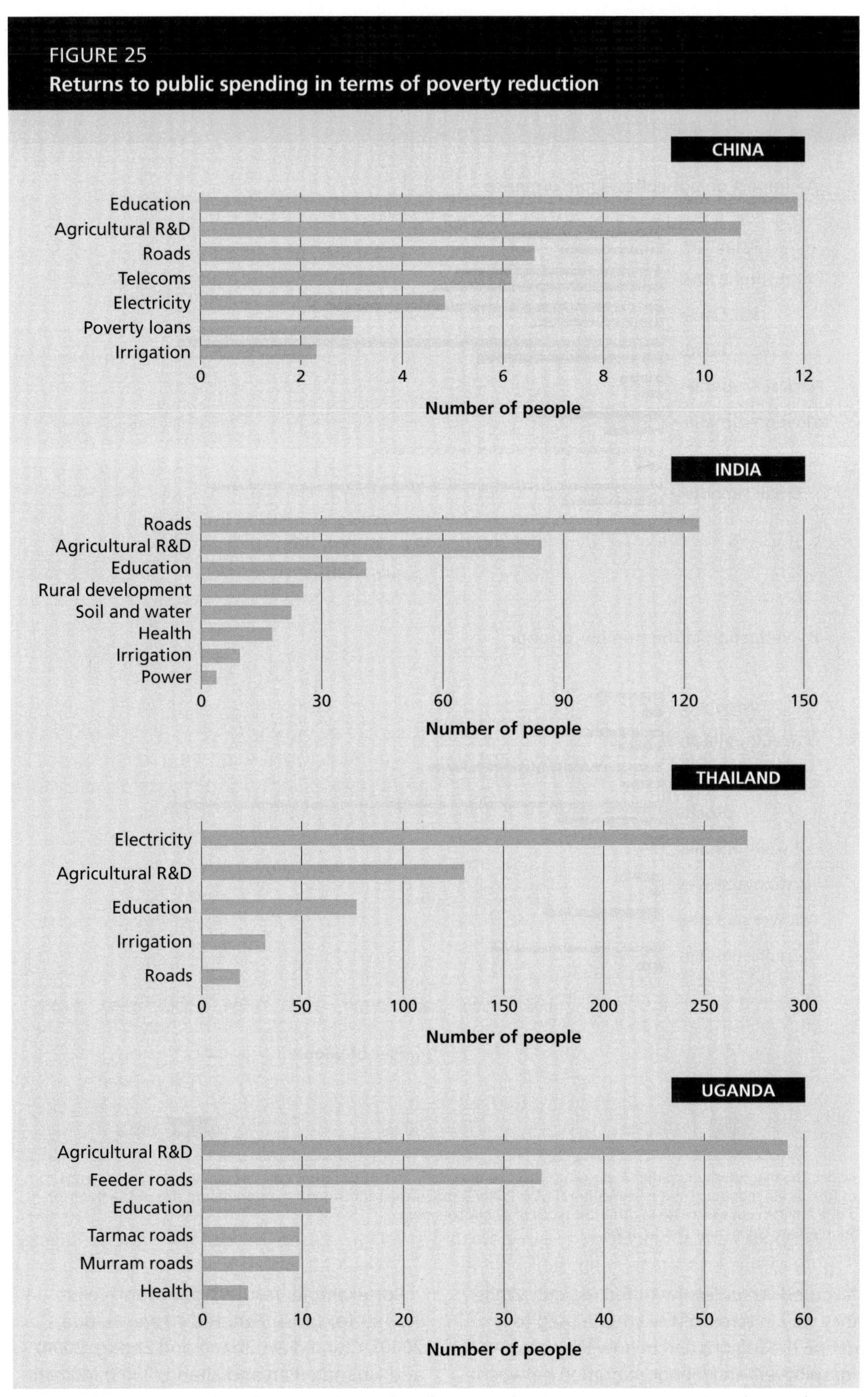

Notes: The magnitudes are the reductions in the number of poor people per monetary unit spent in each area of expenditure. The respective monetary units are: one million baht in Thailand (i.e. reduction of number of poor people per one million baht spent in different sectors); one million rupees in India; 10 000 yuan in China; and one million Ugandan shillings in Uganda.
Sources: Fan, Zhang and Zhang, 2004; Fan, Hazell and Thorat, 2000; Fan, Yu and Jitsuchon, 2008; Fan and Zhang, 2008.

FIGURE 26
Historical impact of various types of public investment and subsidies on agricultural performance and poverty in India

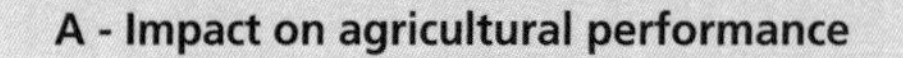
A - Impact on agricultural performance

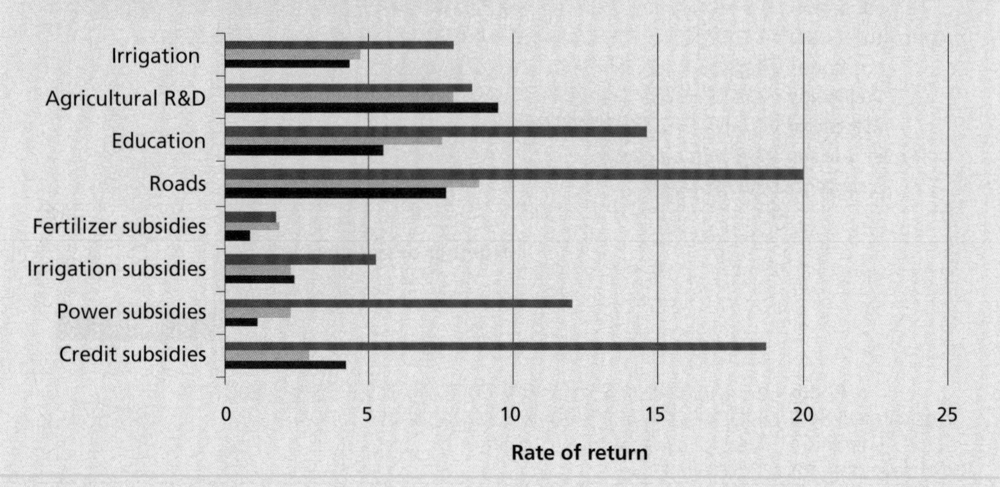

B - Reduction in the number of poor

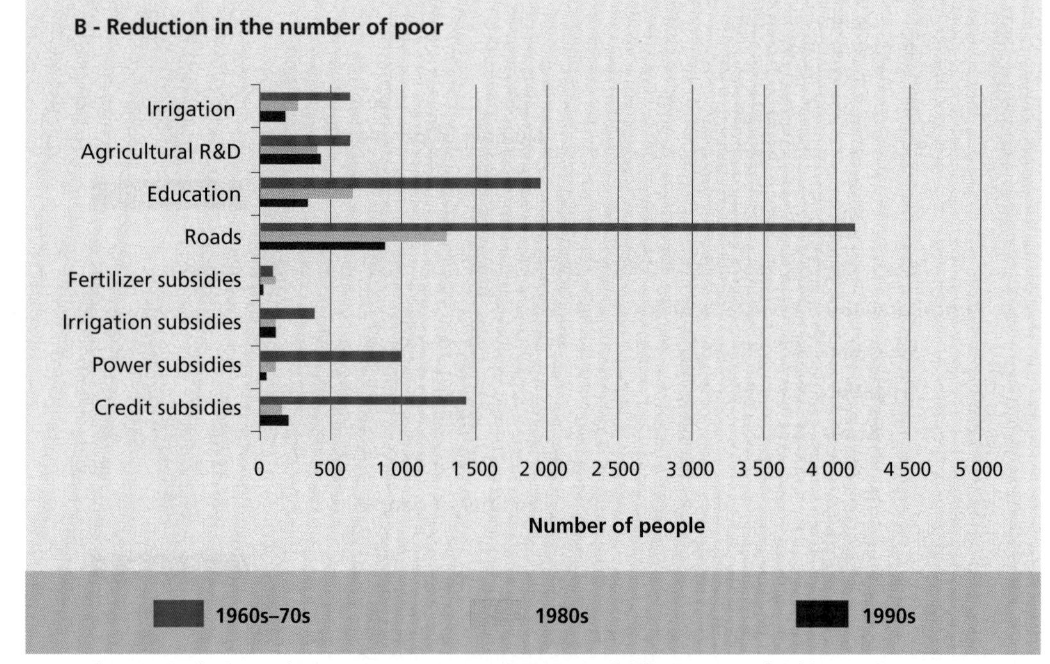

Notes: The magnitudes in panel A are returns to one monetary unit of different types of public spending in terms of (the same) monetary unit of agricultural GDP. Panel B shows the reduction in the population size of the poor for a one million rupee increase in different types of public spending.
Source: Fan, Gulati and Thorat, 2008.

favoured versus less-favoured regions. While they vary in terms of the criteria used to define the regions and in how they account for spillovers and labour migration between regions, the results suggest that public investment in less-favoured regions may have higher returns both in terms of agricultural performance and poverty reduction.

For example, results from countrywide studies for India (Fan, Hazell and Haque, 2000), China (Fan, Zhang and Zhang, 2004), and Uganda (Fan and Zhang, 2008) indicate that investment in less-favoured regions may have higher payoffs. These results are summarized in Figure 27; note that the distinction between high-potential and

TABLE 13
Share of subsidies and public goods in rural government expenditures in Latin Americ and the Caribbean, selected countries

COUNTRY	SUBSIDIES AS SHARE OF RURAL SPENDING	PUBLIC GOODS AS SHARE OF RURAL SPENDING
	(Percentage)	
Argentina	59	41
Brazil	87	13
Costa Rica	48	52
Dominican Republic	80	21
Ecuador	69	31
Guatemala	27	73
Honduras	9	91
Jamaica	58	42
Mexico	66	34
Nicaragua	37	63
Panama	51	49
Paraguay	32	68
Peru	64	37
Uruguay	19	82
Venezuela	54	46

Note: The shares are annual averages for the years 1985 to 2001.
Source: López and Galinato, 2006.

participation, transparency and ethno-linguistic fractionalization. Indeed, higher levels of inequality tended to increase both the overall government allocation to rural areas and the share of subsidies within overall agricultural expenditures. However, the authors emphasized the need for further data collection and analysis to determine whether their conclusions would hold outside the region.

Political economy of public investment in agriculture[26]

If returns to public investment are so high, why don't governments invest more? And if returns to public investment are higher than returns to subsidies, why do governments continue to subsidize? The analysis just reported by Allcott, Lederman and López (2006) pointed to the role of wealth distribution, along with other political and institutional factors, as determinants of the structure of rural public expenditure. The question of how public expenditure policies relating to agriculture are actually determined is important for understanding how to improve public investment.

A fundamental difference between private and public investments decisions is that, while the former are motivated by expectations of private returns, the latter should in principle be motivated by expected social returns. In reality, for a number of reasons, the motivations of decision-makers may not coincide with the wider social benefits expected from the investment. Public expenditure and investment patterns can be affected by factors such as pressures by interest groups, corruption or even the characteristics of agricultural investments themselves. For instance, some agricultural investments may have very long pay-off periods and their impacts may not always be clearly identified, so politicians – who are interested in remaining in office – may not get much credit. Factors as these, as well as governance in general, can have a major impact on how public funds for agricultural expenditures and investment are used.

[26] This section is based on a background paper prepared by an IFPRI staff member.(see Mogues, 2012).

FIGURE 27
Returns to various investment types in high-potential versus less-favoured lands

AGRICULTURAL PERFORMANCE

POVERTY REDUCTION

CHINA

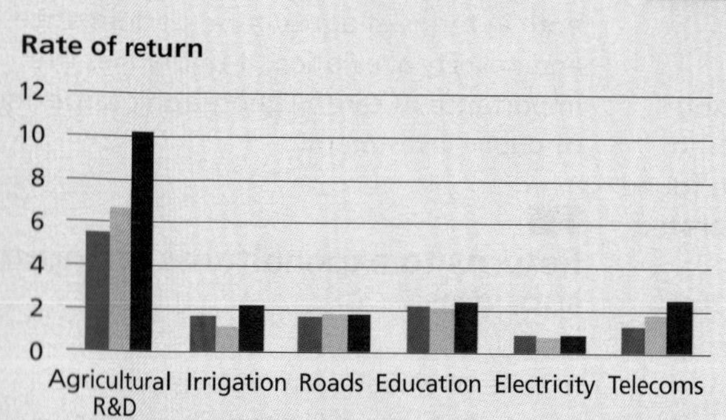

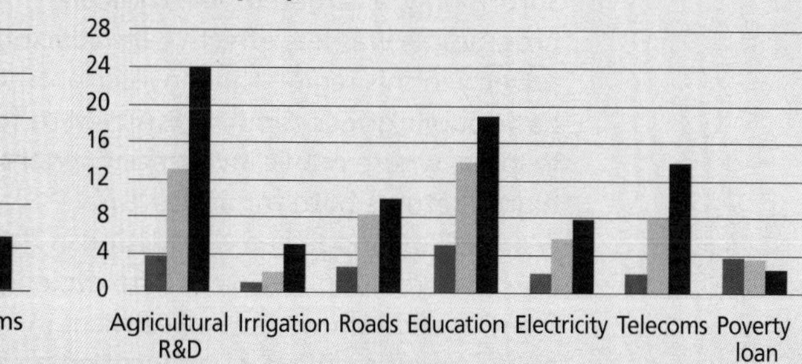

Coastal | Central | Western

INDIA

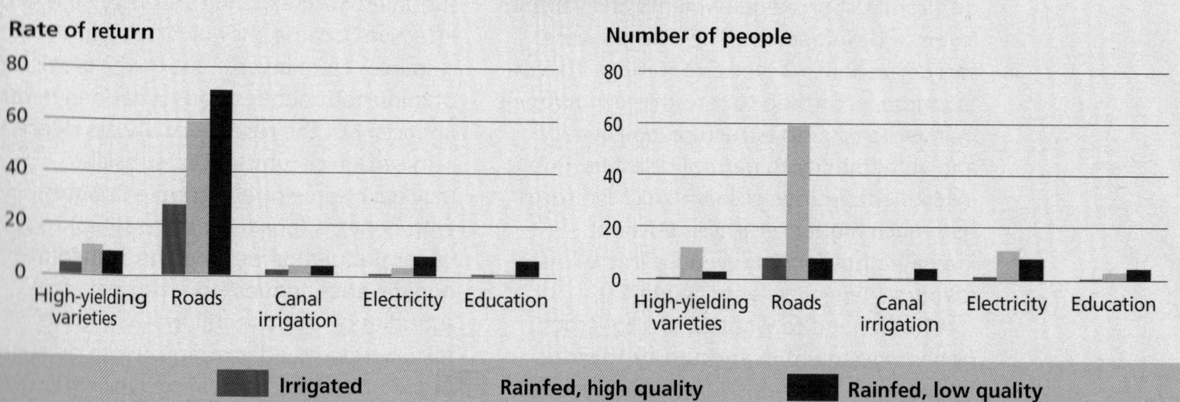

Irrigated | Rainfed, high quality | Rainfed, low quality

UGANDA

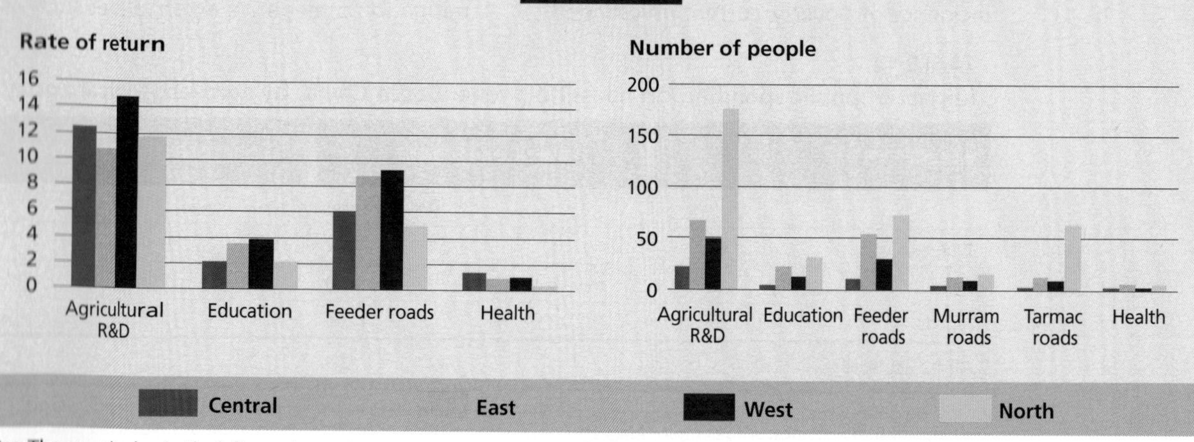

Central | East | West | North

Notes: The magnitudes in the left panel are returns to one monetary unit of different types of public spending in terms of the value of agricultural production or productivity expressed in the same monetary unit. The agricultural performance variable is measured slightly differently in each country: agricultural GDP in China, agricultural total factor productivity in India, and agricultural labour productivity in Uganda. The magnitudes in the right panel are the reductions in the population size of the poor per monetary unit spent in each area of spending. The respective monetary units are: one million rupees in India; 10 000 yuan in China; and one million Ugandan shillings in Uganda.
Source: Fan, Zhang and Zhang, 2004; Fan, Hazell and Haque, 2000; Fan and Zhang, 2008.

less-favoured areas differs for the three countries, but in the graphic the more-favoured areas are found to the left and the less-favoured areas to the right.

In China, investment clearly had the highest returns in the least-favoured western region, both in terms of agricultural performance and poverty reduction. Surprisingly, a targeted poverty-loan programme was less effective in reducing poverty in this region than investments in basic public goods. Similar results were found in India, where public investment generated higher returns both for agricultural productivity and poverty reduction in the less-favoured rain-fed areas. The evidence from Uganda shows that investment in public goods such as R&D, education and roads clearly had a stronger poverty-reducing impact in the less-favoured northern region.

Similar results were found by Dong (2000), who looked at ten Chinese villages with different resource endowments and varying levels of development. The villages were categorized into 3 types from more- to less-favoured, according to resource endowment, market access, infrastructure, soil quality and vulnerability to natural disasters. Public investment and social service expenditures had much higher returns in terms of increased household revenues in the less-favoured Type III villages (Table 12).

Whether and to what extent to target public investment in agriculture to more-favoured or to less-favoured areas remains an empirical question. The answer will likely depend on local circumstances, incidence of poverty, current investment levels and the potential for spillovers and labour migration between the regions. Nevertheless, the limited evidence presented above suggests the existence of situations of underinvestment in less-favoured areas, where redirecting agricultural investment to these areas could generate higher returns both in terms of agricultural performance and poverty alleviation. It underlines the importance of careful geographic targeting of public investment.

Returns to expenditures on input subsidies

In spite of evidence of high returns on investment in public goods in and for agriculture, in practice significant amounts of government expenditures both in developing and developed countries are devoted to current expenditures in the form of subsidies. Such expenditures may be less cost-effective because they divert scarce public resources from investment in the provision of important public goods with longer-term impacts, but the case is not always clear-cut.

In certain circumstances, subsidies may have some public good attributes, with benefits (positive externalities) to a wider population beyond the immediate beneficiaries. Indeed, the rationale for subsidies on agricultural inputs such as fertilizer and seed is often pinned on such arguments. The use of improved agricultural technologies can have economic and social benefits beyond the farm, including the mitigation of negative externalities such as depletion of soil fertility and the expansion of farming into marginal areas. The balance of the evidence on the relative returns to fertilizer subsidies versus investment in more clearly recognized public goods suggests that subsidies may be over-used.

Although returns to subsidies for fertilizers and other inputs appear to decline over time, (Fan, Gulati and Thorat, 2008), their use has increased rapidly in many countries in recent years. Subsidies often rise with a country's fiscal capacity (Byerlee, de Janvry and Sadoulet, 2009). In India, agricultural subsidies rose from 40 percent of agricultural public expenditures in 1975 to 75 percent in 2002, and by 2002/03 accounted for 6 percent of agricultural GDP (World Bank, 2007a). In Indonesia, fertilizer subsidies accounted for 30 percent of total agricultural development spending by 1988-90, although investments in research, extension and irrigation infrastructure were more important drivers of output growth during the previous two decades (Rosegrant, Kasryno and Perez, 1998).

Several countries have initiated fertilizer subsidy programmes in recent years. In Zambia, by 2005, about 37 percent of the agriculture budget was spent on fertilizer subsidies while irrigation development and other rural infrastructure received only 3 percent and agricultural R&D only 0.5 percent (World Bank, 2007a). The budgeted cost of the programme quadrupled from 2002/03 to 2008/09 (World Bank, 2010b), and evidence shows that it has crowded out private suppliers in areas where they had been active (Xu *et al*., 2009).

Malawi also re-introduced universal fertilizer subsidies in 2005/06, and by 2008/09 up to 1.5 million households were expected to receive vouchers for a total of 182 300 tonnes of subsidized fertilizer. The programme successfully raised maize output, but absorbed 16 percent of Malawi's total government budget in 2008/09 and, because fertilizer is distributed by a state company, displaced private sector participation (Wiggins and Brooks, 2010).

There is a significant amount of research on the returns to public expenditures on a range of public goods, but little attention has been devoted to the impact of the overall composition of public expenditures and their breakdown into public and private goods. While the distinction between public and private goods is not always rigidly defined, the allocation of public funds to subsidies for goods such as agricultural inputs that primarily benefit private individuals can divert funds away from public goods and other socially beneficial expenditures. The allocation of public resources to subsidies may thus have significant implications both in terms of economic efficiency and social equity.

Evidence on the efficiency and equity implications of the structure of rural expenditures was analysed for 15 countries in Latin America and the Caribbbean for 1985-2001 by López and Galinato (2006). They classified public expenditures as either public goods or subsidies. The share of subsidies in rural expenditure in this time period ranged from less than 10 percent to almost 90 percent (Table 13).[25]

López and Galinato (2006) found that the overall level of government expenditures in rural areas had a positive and highly significant impact on per capita agricultural GDP, but the composition of government expenditure in terms of subsidies was much more important. Increasing the share of subsidies, while keeping total expenditures constant, significantly reduced per capita agricultural GDP. According to their estimates, just reallocating 10 percent of rural public expenditures from subsidies to public goods would increase per capita agricultural incomes by 5 percent. Also, increasing overall public expenditure on agriculture would have positive growth effects, but they are smaller than those deriving from reallocating within a given overall budget. The key policy message emerging from this analysis is that governments can increase agricultural GDP just by shifting agricultural expenditures from subsidies to public goods.

Additional analysis by Allcott, Lederman and López (2006), based partly on the same dataset, looked at the determinants of the level and composition of rural public expenditures and of agricultural growth. They found that historical wealth inequality was a key determinant, together with other political and institutional factors such as government accountability, civil society

TABLE 12
Impact of public spending on household revenues in China, by agro-ecological zone

VILLAGE TYPE	PUBLIC INVESTMENT	SOCIAL SERVICE EXPENDITURE
	(Estimated marginal rate of return)	
Type I	1.1	1.5
Type II	2.0	2.7
Type III	7.4	8.2
All households	**3.9**	**4.6**

Notes: Household returns are measured as gross revenue of household operations, including both agricultural and non-agricultural activities. Wage employment and other income generating activities outside of household production are excluded. The marginal rates of return are estimated by multiplying gross revenue by the regression coefficients and dividing the product by 100. Public investment includes maintenance of village irrigation networks and roads. Social services include mechanized ploughing, crop protection, threshing, technical guidance, subsidizing farm inputs, marketing assistance, and other non-agricultural services such as installing drinking water, enhancing access to electricity and providing educational services (schools, libraries and day-care).
Source: Dong, 2000.

[25] Note that these results are not comparable with those reported in Box 5 because they come from different sources and use different definitions of public goods.

Government investment may not always lead to expected results because of excessive costs, low rates of return of the asset resulting from the investment or misuse of the asset, once created. Many factors can reduce the efficacy of government investment: bribery can increase the cost of an investment; governments may simply be ineffective in controlling costs; aid financing may lead to the choice of more expensive projects; and decisions may be subject to patronage or political considerations (Pritchett, 1996). One source of misuse of an asset created through public investment can be the lack of provision of funding for operating expenses and maintenance of the asset. The efficacy of government investment, measured as the difference between public expenditures and the value of the assets generated, is closely associated with indicators of good governance and policies (Pritchett, 1996).

Interest groups and collective action

Interest groups can be a strong influence on public expenditure and investment decisions in agriculture. A rich body of evidence has pointed to the ways that agricultural policies in developing countries have tended to favour a small number of larger-scale farmers (see Birner and Resnick, 2010 for a brief overview). Historically, in developing countries, public investment, pricing policies and other measures have benefited the urban population at the expense of rural dwellers and agricultural households (Lipton, 1977). These phenomena have been explained through the characteristics of interest groups, which affect their ability to press for public policies, including investments, subsidies and other public interventions, that are favourable to them (Becker, 1983).

The effectiveness with which different interest groups can influence politicians through collective action depends on several factors (Olson, 1965). Some of these tend to put farmers at a disadvantage relative to urban dwellers. The spatial dispersion of farmers and inferior access to transportation and communication infrastructure makes coordination and mutual monitoring of actions more difficult than for urban citizens (Olson, 1985). Also their larger number in many developing countries puts farmers at a disadvantage relative to urban dwellers. Indeed, for any given level of spatial concentration and access to transport and communication infrastructure, it is harder to coordinate among larger than among smaller groups (Olson, 1965).

A group's influence also depends critically on their financial wealth. This, along with the greater ease of coordination among small groups, explains why a few large farmers can influence public expenditure patterns when wealth and land are highly concentrated (see analysis by Allcott, Lederman and López [2006] cited above). This underlines the importance of increasing the social capital of smaller farmers through producers' associations.

Another significant phenomenon in policy processes involving interest groups is the existence of a status quo bias among policy-makers. Often policies that have outlived their usefulness fail to be discontinued. An example is agricultural input subsidies, which are rarely removed even after they have outlived or failed to meet their initial efficiency-enhancing or equity objectives. Those who benefit from the current state are usually the ones with the requisite power to have ensured policy enactment in the first place (Fernandez and Rodrik, 1991) and who may even see their lobbying power increasing after the policy is already instituted (Coate and Morris, 1999).

Attribution and time lags in benefits

For a policy-maker responsible for decisions on public expenditures, recognition by beneficiaries is likely to be a significant motivation. The ease with which citizens can attribute credit or responsibility to a policy-maker for specific subsidies or investments and their outcomes can therefore have a major influence on the prioritization of public expenditures.

Visible infrastructure projects, such as a school building, or direct transfers are more easily identifiable and attributable to concrete decisions by politicians and officials than, for example, improving the quality of extension services or investing in research and development. The recent surge in large-scale input subsidy programmes can be explained in part by the ease with which impacts can be identified and attributed to the responsible public officials.

The long time lag required before many public investments yield a return makes attribution more difficult. The longer the lag, the more difficult the attribution and the less incentive public officials have to undertake the investment. This is particularly relevant for investment in R&D, which generally has high returns but also a large time lag between the outlays and the benefits. This may represent one of the causes underlying the apparent and systematic underinvestment in R&D discussed above.

The seriousness of the attribution problem also depends on the quality and volume of information and on the level of education of the beneficiaries of the public expenditures. Better -educated citizens with more access to information, mediated for instance by civil society organizations, are better able to make correct attributions. Improving education levels as well as information flows is therefore important for improving the prioritization of public expenditures and investment.

Corruption and rent seeking

Corruption and rent-seeking behaviour can lead to socially sub-optimal patterns of expenditure and investment. Large infrastructure projects easily lend themselves to rent-seeking behaviour by public officials. Evidence from cross-country analysis shows that in low-income countries, the incidence of corruption increases with the share of spending on large-scale capital projects and decreases with the share of social sector spending (de la Croix and Delavallade, 2009).

In countries with high levels of corruption this phenomenon may introduce a bias in favour of large-scale capital projects over other forms of investment or public expenditure. In addition, the pervasiveness of corruption which generates the bias toward large-scale projects is also likely to make those investments less productive than in countries with better governance. Agricultural R&D investments are relatively less prone to rent-seeking and corruptive practices, although there are recorded instances of corruption; for example commodity boards have diverted money from farm levies on farmers that was intended to fund public agricultural research institutes (Omuru and Kingwell, 2006).

Governance and agricultural investment

The governance environment – of which corruption is but one dimension – is increasingly seen as an important determinant of public expenditure allocations, including those for investment in agriculture. Evidence of this causal link supports the strong correlation found between indicators of good governance and the accumulation of on-farm capital stock reported in Figure 16 in Chapter 3.

Deacon (2003) found strong empirical evidence that systems of governance affect the provision of public goods.[27] He found that dictatorial governments consistently underprovided public goods relative to democratic and inclusive governments. He also found that income levels positively affected public goods provision, but that the provision of public goods responded more strongly to income growth in democracies than in dictatorial governments. At the local government level, as well, evidence shows that the share of public investment in total public expenditures of village governments is higher when the village leader is elected rather than appointed (Zhang *et al.*, 2004).

The efficacy of public spending on health and education in achieving the desired outcomes also depends on the quality of governance; such spending in countries with high levels of corruption and inefficient bureaucracy was less effective than in countries with better governance (Rajkumar and Swaroop (2008). Household data from Uganda showed that there was a threshold level of security below which public investment in infrastructure and education had little impact on growth (Zhang, 2004).

Empirical evidence points to a link between different aspects of governance and the provision of public goods by government. The question arises: what are the implications for agriculture and which aspects of governance matter the most for agricultural investment and the provision of public goods? Resnick and Birner (2006) in an overview of empirical evidence on the links between good governance and pro-poor growth discussed the "definitional ambiguity" of governance

[27] Public goods considered were: access to safe water and sanitation, road density, school enrolment and levels of lead in gasoline as an indicator of environmental protection.

and the multiplicity of indicators involved in much of the discussion and the empirical analysis. They pointed to the need for a better understanding of "which aspects of governance are conducive to growth and which determine whether the poor are capable of participating in the growth process" (Resnick and Birner, 2006, p. 38). A similar understanding would seem just as relevant to the specific issue of governance and agricultural investment.

Planning public investment in agriculture

Ensuring more effective public investment in and for agriculture is a major challenge. It involves improving the process of policy-making affecting investment and strengthening planning and budget processes for public investment. The challenge is particularly severe in the low-income and lower-middle-income countries, where agriculture, and especially smallholders, generally plays a central role in economic development and poverty reduction and where resources for investment are more constrained.

There is increasing attention to the need for improving budget processes (see for instance World Bank, 2011e). However, there is also a need to look at policies affecting private investment and at public investment in and for agriculture in an integrated way. Appropriate policies can enhance the returns to both private and public investment. Appropriate public investment can also enhance returns to private investment and improve incentives to invest, but an inappropriate policy framework can significantly reduce their impact and lead to substantial waste of public resources. Many countries are currently making concrete efforts to guide and improve investment in agriculture by developing country investment plans (Boxes 23 and 24).

BOX 23
The Comprehensive Africa Agriculture Development Programme (CAADP)

The Maputo Declaration on Agriculture and Food Security in Africa, adopted in 2003, represents a formal recognition by African countries that the sector is crucial to economic growth and poverty reduction and that greater resources should be devoted to it. In the declaration, the signatory countries committed to a set of principles for promoting agricultural development as well as a clear commitment to specific targets, in particular to allocate at least 10 percent of their national budget to agriculture and to achieve 6 percent annual agricultural growth.

The principles are made operational by the Comprehensive Africa Agriculture Development Programme (CAADP), which provides a common policy framework for agriculture development in Africa. The process involves Country Roundtables to engage with stakeholders, the generation of evidence-based analysis, the development of the investment programmes, assessment and learning from process and practice. These consultations and stocktaking help to distil a consensus among stakeholders about priorities and culminate in the signing of a "Compact", which outlines the country's agenda for agricultural growth, poverty reduction and food and nutrition security. It also specifies responsibilities for the various parties and outlines implementation mechanisms, including coordination and oversight and mobilisation of funding.

The investment plan is then formulated and subjected to a technical review by independent experts to ensure consistency with CAADP principles and objectives, the adoption of best practices,[1] alignment with Compact commitments and operational feasibility of investment programmes. The technical review process is also a condition for qualifying for GAFSP funding (see Box 8 on page 35).

Finally, the High Level Business Meeting is convened by government with participation from national stakeholder

(CONT.)

BOX 23 *(CONT.)*

groups, the CAADP core institutions at national, regional and continental levels, donors and other possible funders. The purpose is to validate and endorse the Investment Plan and confirm implementation readiness and funding commitments as well as agreeing on modalities for implementation.

By March of 2012, 27 countries had signed Compacts, all with Investment Plans ready or being processed, and 19 countries had held the Business Meeting.

Although many challenges remain, there are real benefits to the process. On the positive side, CAADP is helping to foster dialogue and harmonization of agricultural policy-making at the international level. A review of the CAADP framework in Ghana, Kenya and Uganda found that is has been effective at the global and continental levels but that the country-level process was still weak, especially in terms of country ownership, stakeholder participation, use of evidence in decision-making and alignment of policies (Zimmermann *et al.*, 2009). In some cases, funding deadlines (imposed for example by the GAFSP) effectively short-circuited the process of consultations, the evidence-based decision -making the peer review, etc. Donors also did not, at least initially, see the value added in the CAADP process and have generally been slow to respond.

In Rwanda, where CAADP is considered to have been most influential, the government had already previously formulated the Strategic Plan for Agriculture Transformation II (PSTA II), prepared in collaboration with external experts, focusing on identifying potential returns to investment in staple foods and the necessary policy support. The PSTA II was subsequently aligned with the CAADP framework and formed the basis for Rwanda's Investment Plan. The CAADP Secretariat provided technical assistance to identify and cost the PSTA II programmes and sub-programmes. The CAADP-led Business Meeting, i.e. discussion with donors, led to some changes in expenditure priorities. The process has led to more government support and substantial donor pledges, with 80 percent of PSTA II funding now in place.[2]

As Rwanda's PSTA II is a continuation and up-scaling of activities started during the initial PSTA in 2004, it is the only country where sufficient time has elapsed to allow for a tentative assessment of experience with CAADP and investment planning. There have been substantial increases in land use for key staples such as maize, Irish potato, rice and wheat, in part assisted by the mechanization programme, as well as increased adoption of new planting materials and use of fertilizer. Yield increases have been appreciable for all crops.

[1] As suggested in the pillar framework documents, which are a key aspect of the CAADP process and have been developed under the leadership of the Pillar Lead Institutions (see NEPAD, 2010a for more details on the pillars and pillar lead institutions).

[2] On the other hand, the Togo investment plan is funded to the tune of about 10 percent.

Source: Based on Government of Rwanda (2009), NEPAD (2010a), NEPAD (2010b) and NEPAD (2010c).

Key messages

- Public investment in agriculture is strongly supportive of agricultural growth and poverty reduction, but the type of spending matters. Investments in agricultural R&D, rural infrastructure, and education have much higher returns than spending on subsidies for agricultural inputs such as fertilizer. Although the distinction between investment in public goods and subsidies for private goods is not always clear-cut, the evidence from a large number of countries and over a period of 50 years is clear: investing in public goods yields higher returns for agricultural growth and poverty reduction than input subsidies.
- Investments in a broad range of rural public goods are complementary to

BOX 24
The Bangladesh Country Investment Plan

Many low- and middle-income countries in addition to African countries have adopted plans for investment in agriculture. Bangladesh's Country Investment Plan (CIP) – A Road Map toward Investment in Agriculture Food Security and Nutrition – is an example of such and investment planning process.[1] The CIP grew out of the National Food Policy (NFP, approved in 2006) and the related Plan of Action (2008–15) and is built around the three dimensions of food security: availability, access and utilization.

The investment planning process was led by the Government of Bangladesh and involved a wide range of Ministries, Agencies and Departments – with technical, financial and policy support provided by FAO, the United States Department of Agriculture (USAID) and IFPRI. The process involved wide consultations with key ministries, private sector representatives, NGOs, Development Partners and a large number of stakeholders, especially farmers and their organizations.

An important focus of the process has always been the alignment of priorities, thus allowing government agencies and donors to work more effectively towards common goals in line with the principles of the Paris declaration on aid effectiveness (2005). The planning was given impetus by the L'Aquila Food Security Initiative and the US Feed the Future Initiative[2].

Broadly, the Investment Plan aims to: (i) plan and implement investment priorities in a coordinated way; (ii) increase convergence of budget and external sources of funding, and; (iii) mobilize additional resources. Proposed investments relate to strengthening physical, institutional and human capacities in the field of agriculture, water management, fisheries, livestock, agricultural marketing, food management, safety nets, nutrition and food safety.

At a practical level, investment needs are assessed by the various departments that are mandated to contribute to achieving the stated food security goals. Once formulated the projects fall into the government pipeline. The plan incorporates over 400 projects in different areas derived from the NFP Plan of Action (2008–15).

An important aspect of the process is that of monitoring and reviewing the plan. For example, following approval of the first version of the CIP in June 2010, a review process, again involving widespread consultations, was launched in December. An updated version of the CIP was completed in 2011. The intention is for future monitoring and reviewing to generate a successively more refined, more accurately cost assessed, as well as prioritized CIP. In this sense the CIP is thought of as a living document.

[1] The process followed in Bangladesh is very close in spirit and in practice to the CAADP process advocated by NEPAD.
[2] Feed the Future is the United States Government's global hunger and food security initiative. Led by USAID and drawing on the resources and expertise of agencies across the Government, this Presidential Initiative is aimed at helping countries transform their own agricultural sectors to grow enough food sustainably to feed their people.

investments that directly target the agricultural sector; investment in rural roads, for example, tends to improve market access for agricultural producers and encourage private investment in the sector. The relative impact of alternative investments varies by country, suggesting that priorities for investment must be locally determined, but returns to investment in public goods in rural areas are mutually reinforcing.

- Some evidence suggests that investing in less-favoured areas may reduce poverty more effectively than continuing to invest in high-potential areas where significant progress has already been made, but circumstances vary across countries and over time and will depend on the extent to which the impact of investment spreads across regions through technology spillovers, labour migration and economic multipliers.

- In spite of the extensive body of evidence documenting high economic and social returns to investment in public goods that directly and indirectly support agriculture, government budget allocations do not always reflect this priority, and actual spending does not always reflect budget allocations. A number of political economy factors are to blame, including collective action by powerful interest groups, difficulties in attributing responsibility for successful investment activities that have long lead times and diffuse benefits (as for many agricultural and rural public goods) and poor governance and corruption. Strengthening rural institutions and promoting transparency in decision-making can improve the performance of governments and donors in ensuring that scarce public resources are allocated to the most socially beneficial outcomes.

6. A policy framework for better investment in agriculture

No one disputes the importance of investing in agriculture as one of the most effective strategies for fighting hunger and poverty and making the transition to sustainable agriculture. Yet those parts of the world where hunger and poverty are most severe have seen stagnant or negative rates of investment over the past three decades both by farmers and governments. They face ongoing challenges of enhancing equitable productivity growth while dramatically improving the environmental sustainability of the sector.

Farmers are and will remain the largest source of investment in agriculture, which means they must be central to any investment strategy. Focusing only on public investment, official development assistance and foreign domestic investment is therefore not enough. Hundreds of millions of farmers worldwide have shown their willingness to invest in their productive activities often despite adverse conditions. However, too often their investments in agriculture are constrained by an unsupportive policy and institutional environment. Imagine what they could achieve with a supportive enabling environment.

A clear understanding of the incentives and constraints farmers confront in different contexts is required to unlock their potential to invest. The public sector plays an indispensible role in creating and fostering a conducive investment climate within which private investment – primarily by farmers but also other rural entrepreneurs and investors – can thrive and generate socially beneficial outcomes. Governments and donors have a fundamental responsibility in this regard. The elements of a conducive investment climate are well-known, but they remain elusive in many regions. Indeed, in many regions, a large and growing share of public spending for agriculture is not directed towards the most economically or socially beneficial investments. If so much is known about how to improve investment in agriculture, why is so little progress being made?

Creating a conducive investment climate for private investment in agriculture: context matters

The preceding chapters have reviewed the challenges involved in creating a supportive environment for private investment in agriculture. However, the priorities and the importance of the different challenges vary by country and region, depending on context. The overall level of economic development and the role of agriculture in the economy, the extent and depth of rural poverty and hunger, the degree of environmental degradation, the quality of governance and level of institutional capacity all must be taken into consideration. Broadly speaking, countries at different income levels will have different investment priorities and challenges.

High-income countries typically have highly developed and highly capitalized agriculture sectors and a generally favourable enabling environment for agricultural investment. They have the capacity to respond to growing effective demand, through *inter alia* enhanced investment. However, in many countries, incentives to invest in agriculture relative to other sectors are heavily influenced by economic and sectoral policies, in many cases creating a strong bias in favour of agriculture.

From an agricultural investment perspective, a key challenge in these countries is to ensure that economic incentives are not tilted towards (or against agriculture) as a result of policies and to ensure a level playing field for investment in agriculture and other sectors. This may mean reducing high levels of direct government support and protection for the sector. This is critical for ensuring an economically efficient allocation of resources and investment patterns in agriculture, both domestically and at the international level. A further key challenge is to ensure that environmental costs and benefits are reflected in incentives so as to promote the sustainability of production.

Middle-income countries have already reached a certain level of accumulation of capital in agriculture, beyond what characterizes the low-income countries (see below). They tend to also have a relatively more diverse agriculture sector both in terms of products and types of entities operating in the sector. The role of agriculture in poverty alleviation is generally moderate, though differing from country to country. Private investment in these countries comes from a multitude of sources (small-holder on-farm investment, corporate investment, FDI) and flow into different types of activities, ranging from small-scale private commercial farms to large-scale enterprises. Some segments of producers may be disadvantaged in terms of their ability to invest relative to others.

In addition to ensuring a level playing field in terms of economic incentives for investment in agriculture *vis-à-vis* other sectors and the incorporation of environmental costs and benefits into agricultural services, improving the enabling environment for investment is in many cases an important challenge. A key policy challenge in these countries is also to avoid discrimination among different types of investors, with a focus on removing factors that may particularly constrain smaller investors and those in less favourable regions. This is important not just for reasons of equity and fairness, but also to ensure an efficient allocation of investment capital. Special support to help farmers invest in sustainable production methods can also be necessary in many contexts.

The **low-income countries** are very far from realizing the potential of the agriculture sector in terms of productivity, production, income generation and poverty alleviation. For a large number of farmers, enhancing agricultural productivity is a core component of strategies to exit poverty. Building farm-level capital endowments – physical, human and natural capital – is critical to achieving this. Increasing the productive assets of smallholders and enhancing their ability to invest is therefore a cornerstone of poverty alleviation efforts.

Unbiased incentives to invest in agriculture, both *vis-à-vis* other sectors and among different investors within agriculture, are as important as in the previously discussed country categories. In addition, improving the enabling environment for investment in agriculture is an indispensable condition for promoting agricultural investment in a large number of countries. However, this alone cannot ensure adequate levels of capital accumulation. Policies and programmes need to be directed towards overcoming constraints to accumulation of productive assets by smallholders. Specific support to investment in sustainable production methods with long pay-off periods is also likely to be critical to ensure improvements in the sustainability of production. Large-scale investment may contribute to capital formation in agriculture, but is unlikely to present a solution to poverty and food insecurity for large numbers of people and poses serious risks to resource-poor rural people unless properly managed. Policies and programmes need to be in place to ensure that such investments are indeed conducive and not detrimental to food security and poverty alleviation of local populations.

Supporting the conducive investment climate through public investment

A favourable climate for private investment must be supported by public investment. There have been increasing calls for more public investment in agriculture and for enhanced spending on agriculture in general. However, expanding overall expenditures on agriculture may not be a simple proposition. It is therefore important to enhance the impact of scarce public funds for agriculture, based on some core principles.

Focusing scarce funds on investment in public goods

Evidence suggests that in many cases the impact of existing levels of public expenditure on agriculture – in terms of both agricultural production and productivity and poverty reduction – can be enhanced by shifting expenditures from subsidies on private goods towards investment in public goods. For example, credit subsidies typically generate low returns for society, but public investment in strengthening financial institutions can facilitate the provision of better credit services and generate higher returns for society.

Targeted social transfers can generate public good benefits by enabling poor smallholders retain and expand their assets.

Investing in research and development

Substantial evidence on the high social returns to public investment in agricultural research and technology in developing countries suggests, quite unambiguously, that there is clear underinvestment in this area. The impact of R&D public spending on agricultural production or productivity is greater than that of spending on other activities directly related to the sector as well as key investments for agriculture, such as rural infrastructures, education, electrification, health, and telecommunication. Expenditures on productivity enhancing agricultural R&D have also been shown consistently to have a very strong poverty reduction impact.

Choosing judiciously among agricultural investments

Not all types of agricultural investment are equal in terms of their returns. When advocating for more funds to agriculture, it is critical to make distinctions between high and low-payoff activities in terms of productivity, poverty reduction, or other outcomes. When choosing among agricultural investments, a series of points are important to consider.

- While the evidence shows that investments in R&D have consistently high returns and poverty reduction impacts, the pattern for other types of agricultural investments depends on country and context.
- Public investment in certain other sectors can make very significant positive contributions to agricultural performance and poverty alleviation. Key areas in this regard are rural roads and education.
- A careful geographic strategy for investment is needed, as returns to government resources on agricultural development are likely to be highly heterogeneous across space. Specifically, evidence presented in the report suggests that in several instances there may have been underinvestment in less-favoured as compared to high-potential areas.
- Policy-makers and other stakeholders should be aware that benefits from some public types of investment may materialise with a long lag, so that short-term analysis may conceal the economic gains from public investments with long gestation periods.

Improving the policy and planning process for agricultural investment

The principles required for promoting investment in agriculture and channelling it towards activities with higher economic and social return are well-known, but translating these principles into policy action is more difficult. Improving public policies and planning of investment in and for agriculture involves a series of key elements.

Defining the objectives

Effective policy and investment planning for agriculture requires a clear definition of the objectives and identification of how the policies and public investment relate to the overall development strategy of a country. Objectives are country specific and must be developed with effective participation of relevant stakeholders. In broad terms, the relative weight of key objectives such as expanding food supply, alleviating poverty and ensuring environmental sustainability are likely to differ among countries at different stages of development.

Ensuring coherence between policies and public investment planning

Ensuring coherence between public policies and investment planning can enhance their impact and improve the likelihood of meeting objectives effectively and efficiently. This means ensuring that policies and public investments are directed towards the defined objectives and are mutually reinforcing rather than contradictory. If policies and investment plans are not consistent and coherent among each other, the impact of both will be significantly diminished. In the absence of an appropriate policy framework public investment funds risk being wasted.

Improving the empirical base for policies and investment planning and impact analysis

Ensuring coherence and effectiveness of policies and public investment requires a solid base of evidence on their nature and impact.

However, this is not necessarily an easy task. Public expenditure reviews for agriculture can provide a crucial overview of actual patterns of public expenditure allocation as a basis for further improvements.[28] Public expenditure tracking surveys focus on budget implementation and can allow the tracking and measurement of expenditures from allocation to final user and assess the extent to which public funds are actually spent for their intended purpose and identify points of leakage. Understanding the impact of policies on incentives for private investors is equally important.[29] Closely linked to this is the need for capacity development for policy-making at all levels.

Ensuring coordination across sectors, governments, ministries, agencies and development partners

Agricultural investment can contribute to outcomes usually seen as the concern of other sectors and agencies (for example health and nutrition), and investments undertaken by agencies not centrally concerned with agriculture (such as road infrastructure, electrification, education, etc.) can be among the most important contributors to increasing agricultural growth. This points to the need for addressing any administrative and institutional obstacles that hinder coordination across agencies – not only across ministries in developing country governments, but also across units in donor agencies. Also, coordination between different layers of government investing in and for agriculture is important in many contexts. A first (and easier) step may be improving the sharing of information about these types of cross-sectoral effects of public investment, and about the amount and features of investment activities being undertaken by different agencies. A second and more challenging step would be to attempt to improve allocation across and within agencies for mutual benefit and achievement of multiple development goals.

[28] The World Bank and DFID have developed guidelines for the conduct of public expenditure reviews (World Bank, 2011e). IFPRI has conducted a range of studies on returns on different types of public expenditures and investment in different countries (some of them cited in Chapter 5).

[29] The Monitoring and Analyzing Food and Agricultural Policies in Africa Project (see Chapter 3) is one initiative aiming at improving the analysis of both policies and public expenditures.

Improving governance, transparency and inclusiveness in policies and planning

Improving governance, including transparency and inclusiveness, in public policies and investment priorities is crucial to maximizing impact of policies. As an extension of the coordination across sectors and agencies, it is important to ensure the involvement of all relevant stakeholders in defining and implementing policies and investment programmes. Administrative and political decentralization can often contribute to increased transparency and accountability.

Overcoming the political economy constraints

Directing policies and public expenditures towards clear development and poverty reduction objectives is often made difficult by the specific political economy constraints prevalent in different countries and contexts. The main problems are those of avoiding elite capture and of overcoming resistance to change on the part of the beneficiaries of current policies. Overcoming the political economy constraints may be the most difficult hurdle towards improved policies for the promotion of private investment and better public investment in agriculture. However, progress in the areas above – clarification of objectives and development strategies, policy coherence, improvement of the evidence base for policy and investment decisions, better coordination and greater transparency – can contribute towards creating the political support that is necessary for change.

Key messages of the report

The State of Food and Agriculture 2012: Investing in agriculture for a better future offers the following key messages:

- **Investing in agriculture is one of the most effective strategies for reducing poverty and hunger and promoting sustainability.** The regions where agricultural capital per worker and public agricultural spending per worker have stagnated or fallen during the past three decades are also the epicentres of poverty and hunger in the world today. Demand growth over the coming decades will put increasing pressure on the natural resource base, which in many developing regions is already severely degraded. Investment is needed for conservation of natural resources and the transition to sustainable production. Eradicating hunger sustainably will require a significant increase in agricultural investment and, more importantly, it will require improving the quality of this investment.

- **Farmers are by far the largest source of investment in agriculture.** In spite of recent attention to foreign direct investment and official development assistance, and in spite of weak enabling environments faced by many farmers, on-farm investment by farmers themselves dwarfs these sources of investment and also significantly exceeds investments by governments. On-farm investment in agricultural capital stock is more than three times as large as other sources of investment combined.
- **Farmers must therefore be central to any strategy for increasing investment in the sector, but they will not invest adequately unless the public sector provides a favourable climate for agricultural investment.** The basic requirements are well known, but still too often ignored. Poor governance, absence of rule of law, high levels of corruption, insecure property rights, arbitrary trade rules, taxation of agriculture relative to other sectors, failure to provide adequate infrastructure and public services in rural areas and waste of scarce public resources all increase the costs and risks associated with agriculture and drastically reduce incentives for investment in the sector.
- **A favourable investment climate is indispensable for investment in agriculture, but it is not sufficient to allow many smallholders to invest and to ensure that large-scale investment meets socially desirable goals.**
  - **Governments and donors have a special responsibility to help smallholders overcome barriers to savings and investment.** Smallholders often face particularly severe constraints to investing in agriculture because they operate so close to the margins of survival that they are unable to save or to tolerate additional risk. They need more secure property rights and better rural infrastructure and public services. Stronger producers' organizations would help them manage risks and achieve economies of scale in accessing markets. Social safety nets and transfer payments may help them accumulate and retain assets, either in agriculture or in other activities at their choice.
  - **Governments, international organizations, civil society and corporate investors must ensure that large-scale investments in agriculture are socially beneficial and environmentally sustainable.** Large-scale investments, including by foreign corporations and sovereign investors, may offer opportunities for employment and technology transfer in agriculture but may also pose risks to the livelihoods of local populations, especially in cases of unclear property rights. Governance of these investments must be improved by promoting transparency, accountability and inclusive partnership models that do not involve transfer of land and allow local populations to benefit.
- **Governments and donors need to channel scarce public funds towards the provision of essential public goods with high economic and social returns.** Public investment priorities will vary by location and over time; but evidence is clear that some types of spending are better than others. Investments in public goods such as productivity-enhancing agricultural research, rural roads and education have consistently higher payoffs for society than spending on fertilizer subsidies, for example, which are often captured by rural elites and distributed in ways that undermine private input suppliers. Such subsidies may be politically popular, but they are not usually the best use of public funds. By focusing on public goods, including sustainable natural resource management, governments can enhance the impact of public expenditures both in terms of agricultural growth and poverty reduction. Governments must invest in building the institutions and human capacity necessary to support an enabling environment for agricultural investment.

Part II

WORLD FOOD AND AGRICULTURE IN REVIEW

A FOCUS ON PRODUCTIVITY

Part II

A focus on productivity

The current world food and agricultural situation is characterized by continued high and volatile international food prices and the persistence of hunger and malnutrition in many parts of the world. This is generating growing concerns about the long-term sustainability of agricultural and food systems. These problems lie at the heart of recent discussions by the G20 Ministers of Agriculture and the United Nations Conference on Sustainable Development (Rio+20 Summit), both held in June 2012, which emphasized the need for sustainable growth in agricultural productivity to help eradicate hunger and ensure more efficient use of natural resources.

This part of the report examines price trends on international and domestic food markets and reviews recent developments in agricultural production, consumption and trade with a special focus on the supply response to higher food prices. It concludes by discussing the constraints to future output growth and the need for efforts to boost productivity growth in agriculture.

HIGH REAL FOOD PRICES

After declining in real terms throughout the 1980s and 1990s, international food prices began rising in 2002 in an apparent reversal of this long-term trend (Figure 6B on page 16). By 2011, the FAO Food Price Index reached more than double its level during 2000–02 (Figure 28). Perhaps more significant is the fact that real prices have remained above their previous low for more than ten consecutive years. This is the longest sustained cyclical rise in real prices experienced in the last 50 years. While international food prices have come down slightly from their 2011 peak, they still remain well above historical averages and cereal prices increased again in mid-2012.

FIGURE 28
FAO Food Price Index and indices of constituent commodities

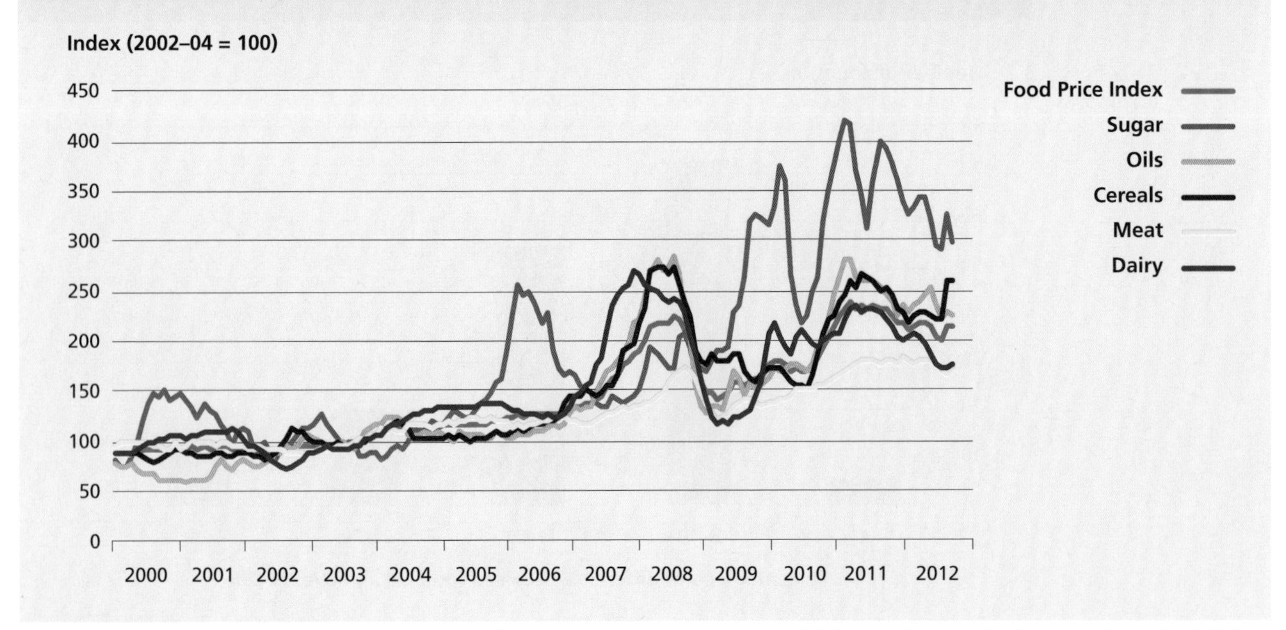

Note: The price indices are monthly observations plotted from January 2000 through August 2012. They reflect actual prices, not adjusted for inflation.
Source: FAO.

Among the commodities that make up the FAO Food Price Index, prices for sugar, oils and cereals showed the sharpest increases in 2010 and early 2011. The volatility of sugar prices has been even more pronounced than that of the other commodities in the index. Meat prices have risen least and have shown less marked fluctuations. Dairy prices have been below the FPI average since late 2010 and have fallen markedly in recent months. International commodity prices are projected in the *OECD-FAO Agricultural Outlook 2012–2021* to remain on a higher plateau for the next decade (OECD-FAO, 2012).

Consumer food prices have risen more rapidly than overall consumer prices since 2000 in all but six of the 166 countries for which data are available (Figure 29). Food price inflation exceeded overall consumer price inflation by up to 10 percentage points in 73 countries, up to 20 percentage points in 55 countries and more than 30 percentage points in 12 countries. Selected country examples illustrate that food price inflation has been particularly severe in countries such as China, Rwanda and Thailand (Figure 30).

The shift towards higher and more volatile agricultural commodity prices can be explained by many factors including, *inter alia*, population growth and higher per capita incomes, urban migration and associated changing diets in developing countries, weather-related production shocks, trade policy shocks and rising demand for biofuel feedstocks (OECD-FAO, 2012). The role of speculative trading as a factor underlying price volatility has also been debated. These factors, combined with tighter natural resource constraints, raise questions regarding the capacity of global agriculture to keep pace with growth in demand. How has global production responded to price trends, and how may it evolve in the future? Which countries have responded most to greater incentives provided by higher commodity prices? How has consumption been affected? Are new trading patterns emerging?

TRENDS IN AGRICULTURAL PRODUCTION, CONSUMPTION AND TRADE

Agricultural production responses

Global agricultural production growth declined somewhat from the 1960s through the 1980s before resuming higher rates of growth in recent years (Table 14). This

FIGURE 29
Average difference between food prices and overall consumer prices, 2000–11

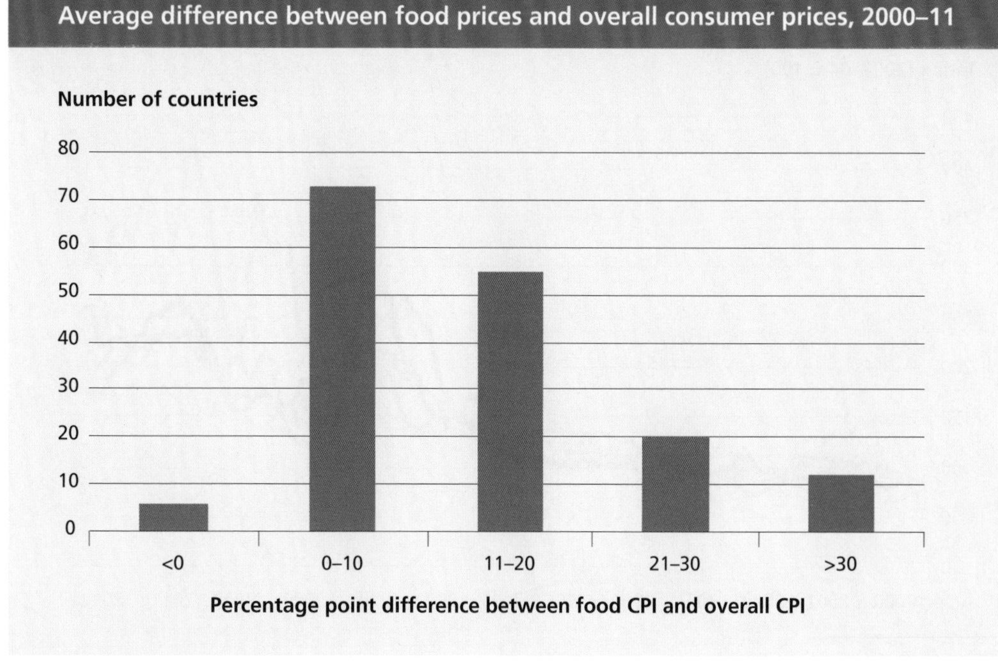

Note: CPI = Consumer Price Index.
Source: FAO, 2012a.

FIGURE 30
Consumer food prices relative to all prices, selected countries

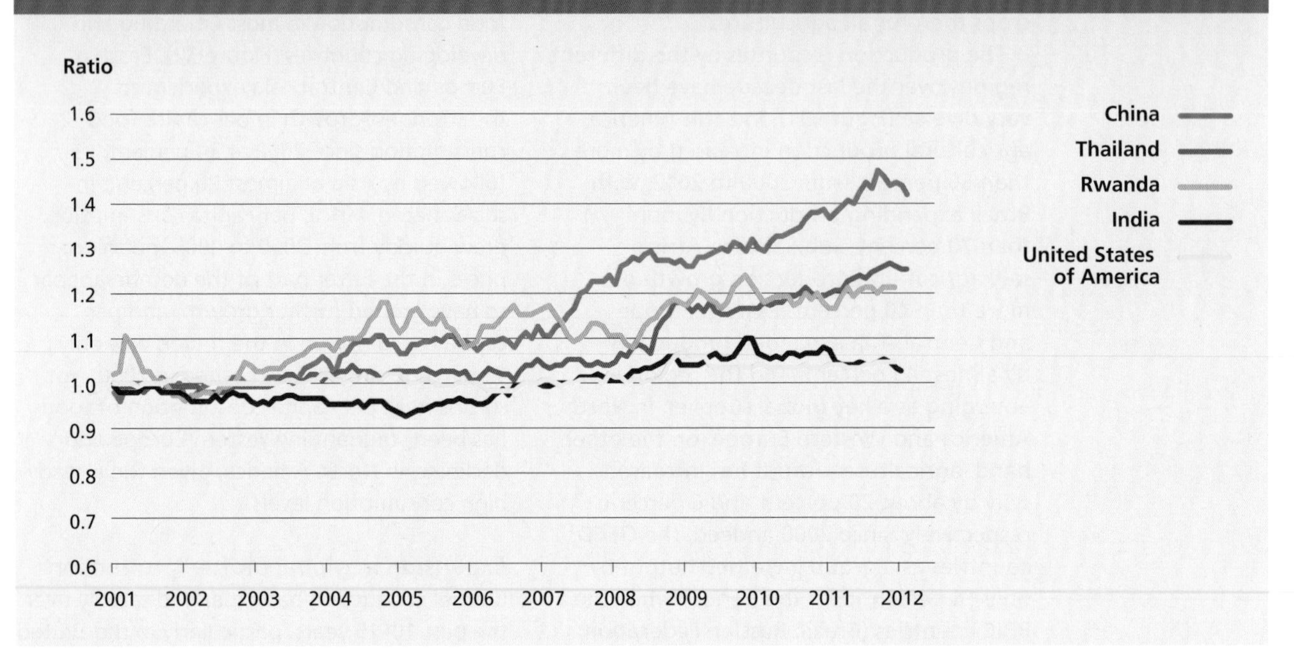

Note: The data on monthly CPI are shown for the period January 2001 through February 2012 for China and Thailand, January 2012 for India and December 2011 for Rwanda and the United States of America.
Source: FAO.

TABLE 14
Average annual growth in agricultural production

	1961–1970	1971–1980	1981–1990	1991–2000	2001-2010
			(Percentage)		
All agriculture					
Total production	2.7	2.4	2.3	2.5	2.6
Per capita production	0.7	0.6	0.6	1.0	1.4
Crops					
Total production	2.7	2.4	2.3	2.5	2.6
Per capita production	0.9	0.4	0.3	1.1	1.5
Livestock					
Total production	2.9	2.5	2.4	2.2	2.2
Per capita production	0.9	0.6	0.7	0.7	1.0

Note: Annual average change in index of net agricultural production. Net production is gross production of crops and livestock net of feed and seed evaluated at 2004-06 constant international reference prices.
Source: FAO.

pattern broadly reflects the long-term price trends discussed above, with the acceleration of production growth in the most recent decade being at least partially attributable to higher price incentives. Total production growth for crops largely mirrors that for all agriculture, whereas total production growth for livestock has not increased in the most recent period, perhaps because prices for livestock products have not risen as much as for crops.

In per capita terms, growth in agricultural production declined very slightly in the latter decades of the last century before

accelerating significantly since 2000. The decline and subsequent recovery of per capita production was more pronounced for crops than for all agriculture.

The production responses by the different regions over the last decade have been very diverse (Figure 31). In Latin America, agricultural production increased by more than 50 percent from 2000 to 2012, with Brazil expanding production by more than 70 percent. Sub-Saharan Africa saw agricultural production growth of more than 40 percent. Eastern Europe and Central Asia expanded production by almost 40 percent, and the region is emerging as a key global supplier. In North America and Western Europe, on the other hand, agricultural output has increased only by about 20 percent and 6 percent, respectively, since 2000. Indeed, the OECD countries as a group increased output by only 14 percent over the period, while the BRIC countries (Brazil, Russian Federation, India and China) increased it by 39 percent, the least-developed countries by 54 percent and the remaining developing countries by 45 percent.

Food consumption

Despite higher prices, rapid income growth has supported robust increases in per capita food consumption in most emerging and developing countries (Figure 32). Eastern Europe and Central Asia experienced the strongest growth in per capita food consumption since 2000 at 24 percent, followed by Asia at almost 20 percent. In sub-Saharan Africa, per capita consumption grew quickly from 2000 to 2005, but higher prices in the latter part of the decade appear to have limited further growth, and per capita consumption in the region was only 11 percent higher in 2012 than in 2000. Not surprisingly, per capita consumption of food has been stagnant in Western Europe and declining in North America, given the already high consumption levels.

Expansion in global biofuel production

Biofuel production has expanded rapidly over the past 10–15 years, particularly in the United States of America, Brazil and the European Union (EU). Ethanol production in the United States of America and Brazil grew by 780 percent and 140 percent respectively over the

FIGURE 31
Net production by region

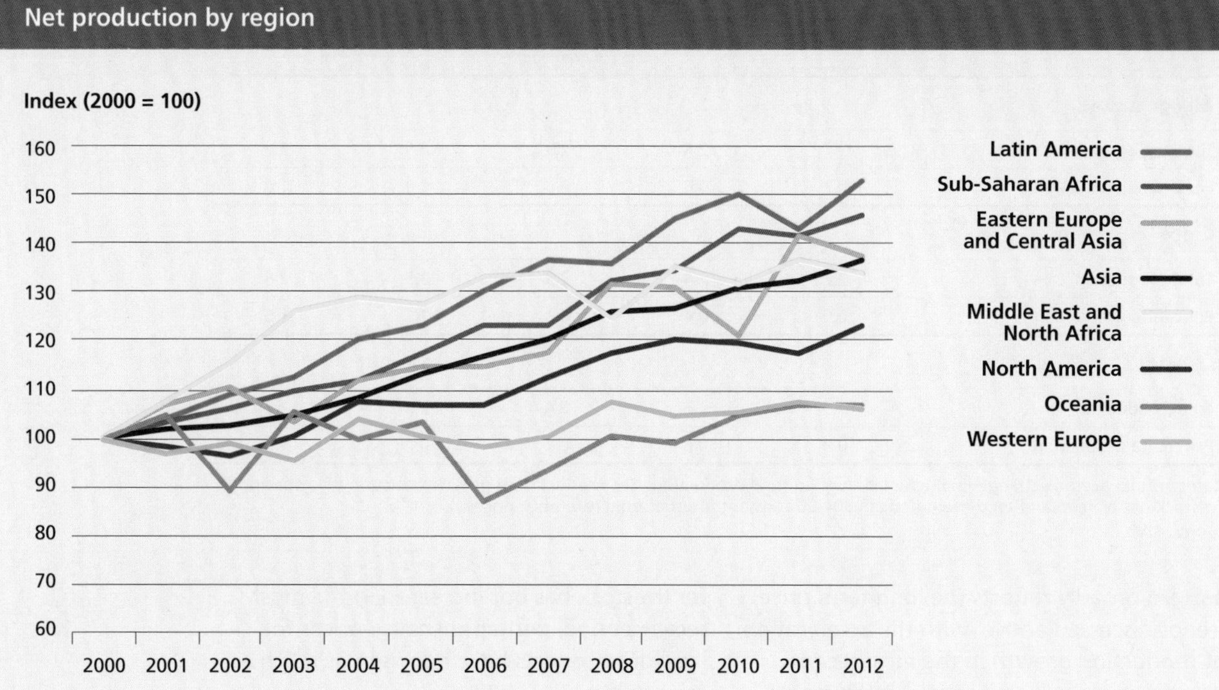

Notes: Net production is gross production of crops and livestock net of feed and seed evaluated at 2004–06 constant international reference prices. Data for 2012 are projections; those for 2011 are provisional estimates.
Source: FAO.

FIGURE 32
Per capita food consumption by region

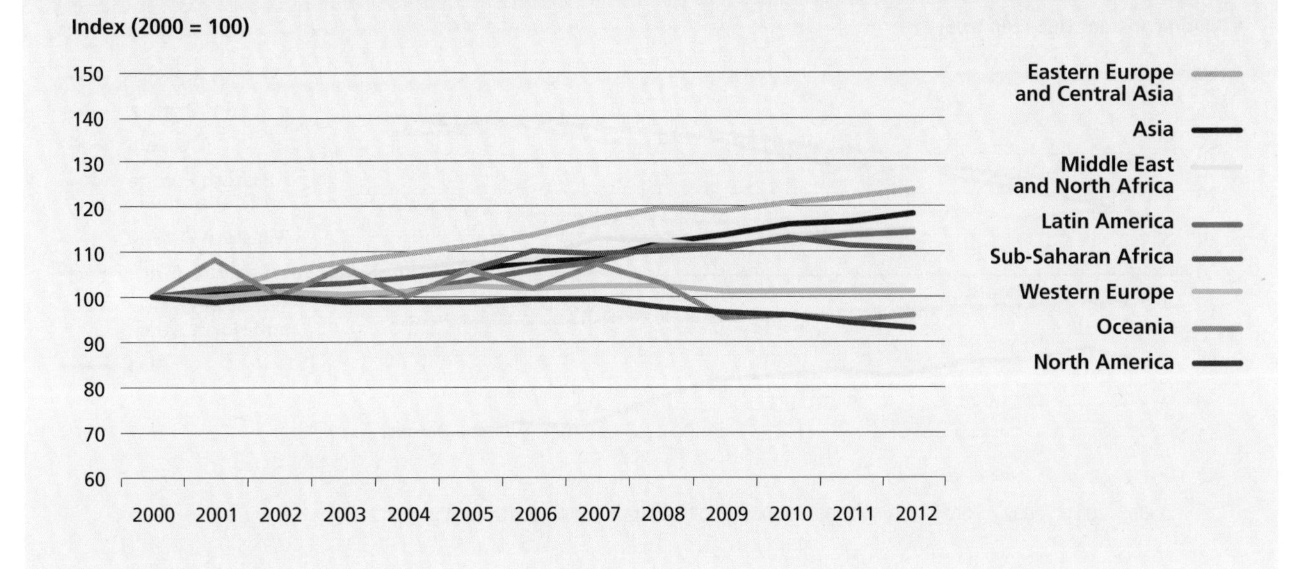

Notes: Food consumption of crops and livestock evaluated at 2004–06 constant international reference prices. Data for 2012 are projections; those for 2011 are provisional estimates.
Source: FAO.

period 2000–12. By 2012, ethanol production absorbed over 50 percent of Brazil's sugar cane crop and 37 percent of the coarse grain crop in the United States of America. Biodiesel production absorbed almost 80 percent of the EU vegetable oil production. In other countries, such as Australia and Canada, growth in the biofuel sector has been strong, although less than in the primary producing countries. Growth of the biofuel sector has been driven largely by policies – such as mandates, blending credits or subsidies and various supportive trade policies – although higher petroleum prices have played a clear role in stimulating demand. The sector has proved the largest source of new demand for agricultural production in the past decade, and represents a new "market fundamental" that is affecting prices for all cereals (de Gorter and Just, 2010).

Changes in global trade patterns

Global trade patterns have changed significantly since 2000 in ways that reflect the underlying trends in production and consumption (Figure 33). The growth of net trade (exports minus imports, in constant dollars) in Latin America has been the strongest of any region, as a result of its significant production growth and in spite of its sustained consumption growth. However, for products considered in this analysis, North America remains the largest net exporter, owing primarily to stagnant consumption in the region. Eastern Europe and Central Asia appears to be moving from a net-importing to a net-exporting region, while Western Europe's trade pattern remains stable as a net importer. Sub-Saharan Africa's net imports continue to grow gradually as high population growth outpaces that of domestic food supply. The Middle East and North Africa is becoming an important and rapidly growing net-importing region, as agricultural production is not keeping pace with demand. However, the most rapidly growing net importer is the rest of Asia, and in particular China.

FUTURE PROSPECTS AND CHALLENGES

The major conclusion from this assessment is that global agriculture appears to be facing a demand-driven expansion supplied primarily by new and emerging exporters rather than traditional suppliers. However, higher input costs and the higher costs of

FIGURE 33
Net exports of food by region

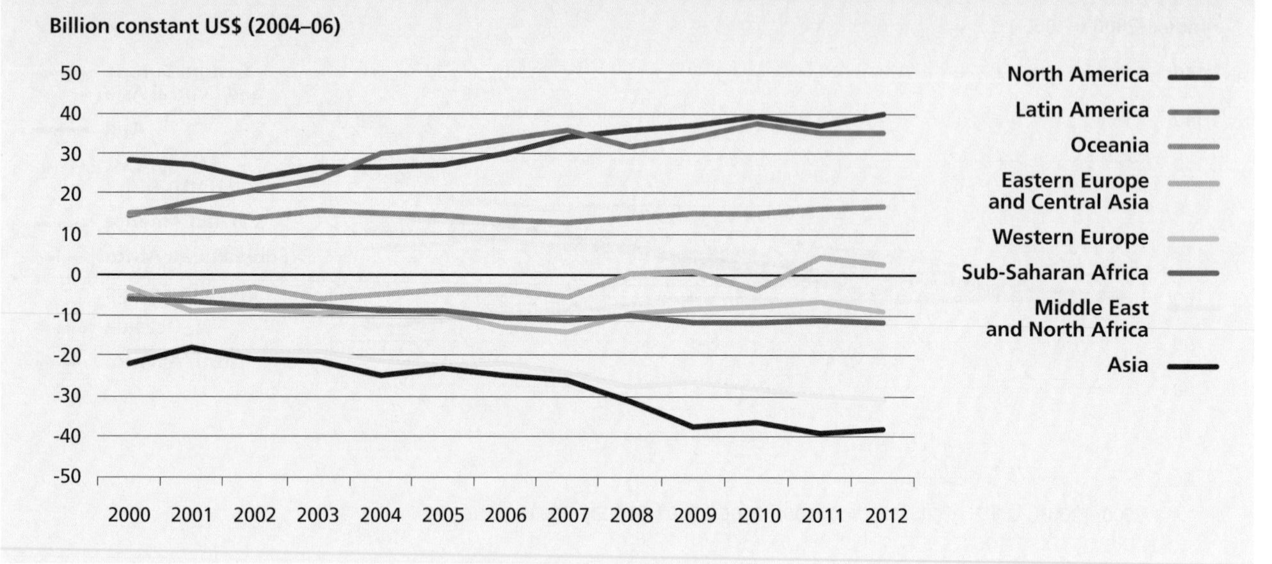

Notes: Net exports of crops and livestock evaluated at 2004–06 constant international reference prices. Data for 2012 are projections; those for 2011 are provisional estimates.
Source: FAO.

access from more remote areas have been driving food prices upwards in real terms. The question is whether production will keep pace with demand in the coming years, so as to either stabilize real prices or bring them down to historical trends, or whether prices will continue to rise under growing demand pressures.

As argued in the *OECD-FAO Agricultural Outlook 2012–21* (OECD-FAO, 2012), food prices are expected to remain on their higher plateau for the next decade. Furthermore, according to the *Outlook* (based largely on the views of national experts and commodity experts at OECD and FAO, as well as on assumptions of "normal" growing conditions, firm economic growth in developing regions and rising real energy prices), the average annual growth in global agricultural production through 2021 will slow to 1.7 percent, down from the 2.6 percent of the previous decade. Agriculture in many countries has grown at a pace that cannot be sustained. Rising input costs and potential supply constraints appear on the immediate horizon. These derive from the availability and quality of resource inputs and the prospects for sustainable productivity growth.

Resource constraints

Globally, most of the best land is already being used in agriculture. Analysis of global agro-ecological zones data reveals that much of the additional arable land is in Latin America and sub-Saharan Africa but is in remote locations, far from population centres and agricultural infrastructure and cannot be brought into production without investments in infrastructure development. Where the potential to expand agricultural land use exists, there is also competition from urban growth, industrial development, environmental reserves and recreational uses, while other areas are not readily accessible or are of poorer quality (FAO, 2011h).

A recent FAO report warns of "the creeping degradation of the land and water systems that provide for global food security and rural livelihoods" (FAO, 2011h). Approximately 25 percent of the world's agricultural land area is highly degraded. These pressures have reached critical levels in some areas, and climate change is expected to worsen the situation (IPCC, 2012; Easterling *et al.*, 2007). There are also other serious resource constraints, especially concerning water. At present, agriculture accounts for over 70 percent of global

TABLE 15
Total factor productivity growth in agriculture, selected regions and countries

	AVERAGE ANNUAL GROWTH RATE				
	1961–1970	**1971–1980**	**1981–1990**	**1991–2000**	**2001–2009**
All developed countries	**0.99**	**1.64**	**1.36**	**2.23**	**2.44**
All developing countries	**0.69**	**0.93**	**1.12**	**2.22**	**2.21**
North Africa	1.32	0.48	3.09	2.03	3.04
Sub-Saharan Africa	0.17	-0.05	0.76	0.99	0.51
Latin America and the Caribbean	0.84	1.21	0.99	2.30	2.74
Brazil	0.19	0.53	3.02	2.61	4.04
Asia	0.91	1.17	1.42	2.73	2.78
China	0.93	0.60	1.69	4.16	2.83
Transition countries	0.57	-0.11	0.58	0.78	2.28
Russian Federation	0.88	-1.35	0.85	1.42	4.29

Source: Fuglie, 2012.

water use, but the share of water available for agriculture is expected to decline to 40 percent by 2050 (OECD, 2012b). The availability of freshwater resources shows a similar picture to that of land: sufficient resources at the global level are unevenly distributed and an increasing number of countries, or parts of countries, are reaching critical levels of water scarcity. Many of the water-scarce countries in the Near East and North Africa and in South Asia also lack land resources. Due to their vulnerability, coastal areas, the Mediterranean basin, the Near East and North African countries and dry Central Asia appear as locations where investment in water management techniques should be considered a priority when promoting agricultural productivity growth.

Prospects for productivity growth

Several studies point to slowing productivity growth in agriculture. For crops, for instance, some evidence suggests a slowdown in yield growth rates in recent decades. The 2008 *World Development Report* (World Bank, 2007) highlighted the decline in annual average yield growth rates for maize, wheat, rice and soybeans, both globally and for most country groupings, with the exception of Eastern Europe for wheat and soybeans. Alston, Beddow and Pardey (2010) reported similar results for developing and developed countries – in particular for cereal yields – in the majority of large producing countries.

While certain measures of partial productivity growth, such as crop yields, may be slowing in some regions, total factor productivity (TFP)[30] growth does not appear to be slowing (Table 15). Indeed, estimates show recent annual growth in TFP in the 2.2–2.5 percent range in both developed and developing regions.

One of the salient characteristics of both partial and total productivity measures is the large differences in absolute productivity among countries. While growth rates may be similar or higher, productivity in developing regions is often a fraction of that in developed regions. Many developing regions also have large gaps relative to their potential. In sub-Saharan Africa, for example, crop yields reached only about 27 percent of their economic potential in 2005 (Figure 34). Closing these yield gaps – by, *inter alia*, providing female farmers and other smallholders with equal access to productive resources – could have a significant impact on crop supply, both regionally and globally, and hence on market balances and commodity prices.

Simulation experiments with the Aglink-Cosimo model employed in the *OECD-FAO Agricultural Outlook* (OECD-FAO, 2012)

[30] Growth in TFP represents that part of production growth that cannot be attributed to increased use of inputs and factors of production but rather by other things such as technological progress, human capital development, improvements in physical infrastructure etc. See also Box 7, where TFP growth is defined and discussed in more detail.

FIGURE 34
Ratio of crop yield to economic potential yield

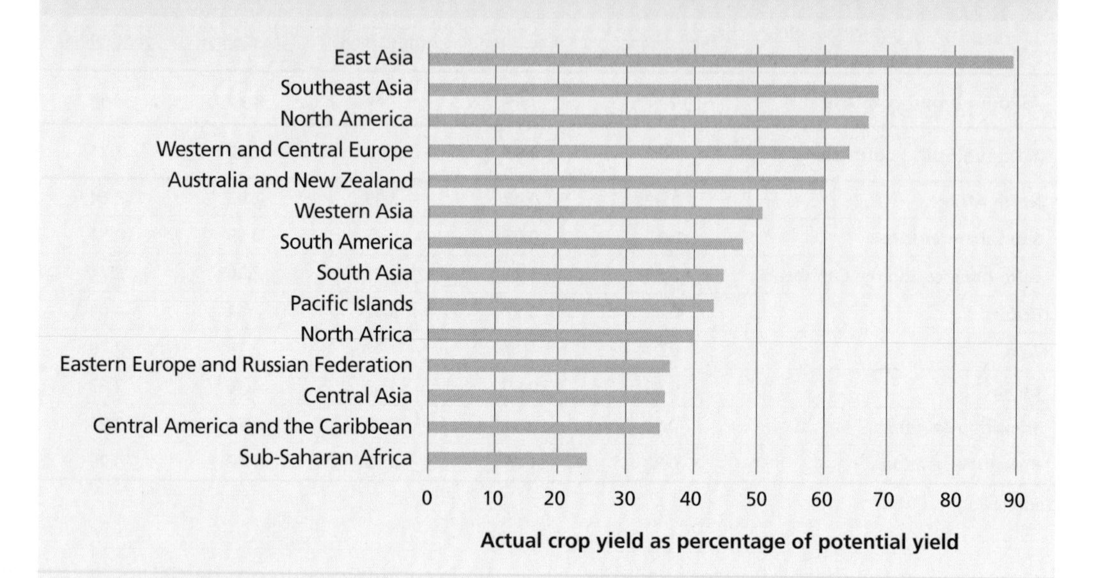

Source: FAO, 2011h.

suggest that reducing cereal yield gaps in developing countries by just 10 percent would increase global cereal supply by about 1.3 percent, 1.8 percent and 2.6 percent for wheat, coarse grain and rice, respectively. Such production increases would lower international prices by 13, 14 and 27 percent, respectively, for each of these commodities. Closing the yield gaps could thus have a considerable impact on agricultural markets and prices.

Reducing food losses and waste is another way to increase food supplies. Global food losses and waste are estimated at roughly 30 percent for cereals; 40–50 percent for root crops, fruits and vegetables; 20 percent for oil seeds; and 30 percent for fish (FAO, 2011i). Food losses occur in both high- and low-income countries. In middle- and high-income countries, food is largely wasted at the consumption stage, whereas in low-income countries it is lost mostly during the early and middle stages of the food supply chain. Investing in more efficient systems that reduce losses or waste would also help to reduce greenhouse gas emissions – both directly, as wastage typically generates methane emissions during food disposal, and indirectly, through the need for fewer resources.

In 2012, at the request of the G20, a number of international organizations jointly prepared a special report on *Sustainable agricultural productivity growth and bridging the gap for small family farms* (Bioversity *et al.*, 2012). This is a clear illustration of the importance governments place on enhancing productivity growth, particularly of smallholder farms. The study assesses the challenges of increasing production and calls on governments to step up their efforts to improve sustainable productivity growth in agriculture by encouraging better agronomic practices, creating the right commercial environment and strengthening innovation systems.

CONCLUSION

The persistence of high levels of undernourishment worldwide and recent trends in agricultural prices, production and consumption confirm the major challenges facing world agriculture over the coming decades, notably meeting increasing demand from a growing world population, contributing to eradicating hunger and malnutrition, and preserving the natural resources upon which agriculture and we all depend. If we are to meet these challenges we need to boost productivity growth in agriculture. Ensuring more and better investments in agriculture is a cornerstone in these efforts.

Part III

STATISTICAL ANNEX

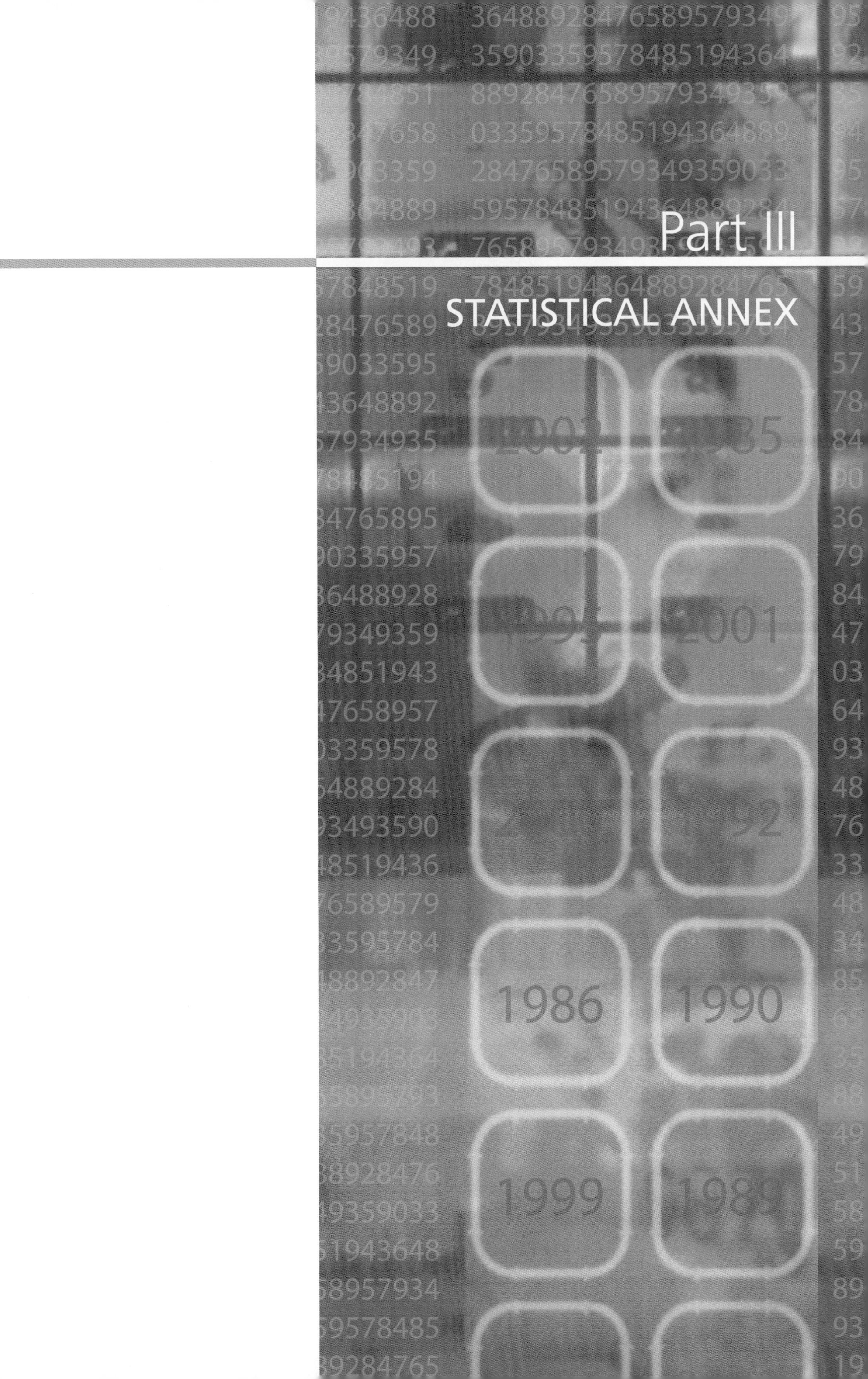

Part III

Notes on the annex tables

Key

The following conventions are used in the tables:

..	= data not available
0 or 0.0	= nil or negligible
blank cell	= not applicable

Numbers presented in the tables may differ from the original data sources because of rounding or data processing. With the exception of Table A3, observations presented in the annex tables include only those observations used for compiling figures and tables in the text. To separate decimals from whole numbers a full point (.) is used.

Weighted averages for income and regional groupings are reported only when data are available for at least half of the countries in each region and represent at least two-thirds of the population of each region.

Technical notes

Table A1. Economically active population in agriculture and agricultural share of total economically active population, 1980, 1990, 2000 and 2010

Source: FAO, 2012a.

The total economically active population includes all employed and unemployed persons. The term covers employers, self-employed workers, salaried employees, wage earners, unpaid workers assisting in a family, farm or business operation, members of producers' cooperatives and members of the armed forces.

Economically active population in agriculture

The number of people engaged in or seeking work in agriculture, hunting, fishing or forestry. Referred to elsewhere in the text as the agricultural labour force or agricultural workers.

Agricultural share of total economically active population

The total number of people economically active in agriculture divided by the total economically active population multiplied by 100.

Table A2. Agricultural capital stock: total and per worker, 1980, 1990, 2000 and 2007

Source: FAO, 2012a.

Agricultural capital stock

Agricultural capital stock equals the total value of a producer's holdings of a defined set of fixed assets. Fixed assets consist of tangible or intangible assets that are used repeatedly or continuously in other processes of production over periods of one year or longer. The physical assets included are land development, livestock, machinery and equipment, plantation crops (trees, vines and shrubs yielding repeated products) and structures for livestock. Values are presented in constant 2005 US dollars.

Agricultural capital stock per worker

Agricultural capital stock divided by the economically active population in agriculture. Values are presented in constant 2005 US dollars.

Table A3. Average annual foreign direct investment inflows to agriculture, food, beverages and tobacco, and all sectors, 2005–06 and 2007–08

Source: Data provided by UNCTAD.

Foreign direct investment (FDI) occurs when an enterprise (the direct investor) establishes a lasting interest in an enterprise (direct investment enterprise) that is a resident of a country other than that of the direct investor. Lasting interest implies the existence of a long-term relationship between the investor and enterprise, as well as the investor's ownership of at least 10 percent of the voting power of the enterprise. FDI flows with a negative sign indicate that at least one of the three components of FDI (equity capital, reinvested earnings or intra-company loans) is negative and not offset by positive amounts of the remaining components. These are instances of reverse investment or disinvestment. Sectoral FDI data use the categories of economic activity established by the United Nations International Standard Industrial Classification of All Economic Activities, Revision 3. Values are presented in current US dollars.

FDI inflows to agriculture

FDI inflows to agriculture are those investments made in crop production, market gardening and horticulture; livestock; mixed crops and livestock; agricultural and animal husbandry services (excluding veterinary activities); hunting, trapping and game propagation; forestry and logging; and fishing, fish hatcheries and fish farms.

FDI inflows to food, beverages and tobacco

FDI inflows to food, beverages and tobacco consist of all such investments in production, processing and preservation of meat, fish, fruit, vegetables, oils and fats; the manufacture of dairy products, grain mill products, starches and starch products, prepared animal feeds, other food products, beverages and tobacco products.

FDI inflows to all sectors

FDI inflows to all sectors are investments to all economic activities. These are: agriculture, hunting, forestry and fishing; mining and quarrying; manufacturing; electricity, gas and water supply; construction; wholesale and retail trade, repair of motor vehicles, motorcycles and personal and household goods; hotels and restaurants; transport, storage and communications; financial intermediation; real estate, renting and business activities; public administration and defence; compulsory social security; education;

health and social work; other community, social and personal service activities; private households with employed persons; and extra-territorial organizations and bodies.

Table A4. Government expenditures: total spent on agriculture and agricultural share of total expenditures, 1980, 1990, 2000 and 2007

Source: IFPRI, 2010.

Total government expenditures are spending carried out by domestic government; they include, as far as possible, the categories considered by the IMF (2001), which are agriculture, defence, education, health, social protection, transportation and communication and others. Values are presented in constant 2005 purchasing power parity (PPP) dollars.

Government expenditures on agriculture

Government expenditures on agriculture include projects and programmes related to administration, supervision and regulation of agriculture; agrarian reform, agricultural land settlement, development and expansion; flood control and irrigation; farm price and income stabilization programmes; extension, veterinary, pest control, crop inspection and crop grading services; production and dissemination of general and technical information on agriculture; and compensation, grants, loans or subsidies to farmers. Expenditures on agricultural research and development as well as on development projects and programmes that serve multiple purposes, including agricultural development, are excluded.

Agricultural share of total expenditures

Government spending on agriculture divided by government spending on all sectors multiplied by 100.

Table A5. Government expenditures on agriculture: per agricultural worker and Agricultural Orientation Index, 1980, 1990, 2000 and 2007

Sources: IFPRI, 2010 and World Bank, 2012.

Government expenditures on agriculture per agricultural worker

Government expenditures on agriculture divided by the total economically active population in agriculture. Values are presented in constant 2005 PPP dollars.

Agricultural orientation index for government expenditures

Agricultural share of government spending divided by the agricultural share of gross domestic product.

Table A6. Public expenditures on agricultural research and development: total and as a share of agricultural GDP, 1981, 1990, 2000 and latest year

Sources: IFPRI, 2012a and World Bank, 2012.

Public expenditures on agricultural research and development

Includes spending by the public sector (government agencies, institutions of higher education and non profit agencies) on research regarding crops, livestock, forestry, fisheries, natural resources and socioeconomic aspects of primary agricultural production as well as on-farm post-harvest activities and food-processing. Values are presented in constant 2005 PPP dollars.

Public expenditures on agricultural research and development as a share of agricultural GDP

Public expenditures on agricultural research and development divided by the agricultural GDP multiplied by 100.

The latest year varies by region. For countries in East Asia and the Pacific the latest year is 2003, with the exception of China for which it is 2008. For those in Europe and Central Asia it is 2000; for Latin America and the Caribbean it is 2006; for the Middle East and North Africa it is 2004; for South Asia it is 2009 and for sub-Saharan Africa it is 2008.

Table A7. Official development assistance to agriculture and agricultural share of ODA to all sectors, 1980, 1990, 2000 and 2010

Source: OECD, 2012a.

Official development assistance (ODA) as presented here consists of commitments of financing made by donor country governments and by multilateral organizations to a recipient country. Such commitments are intended to promote the economic and social development primarily of low- and middle-income countries and are concessional in character with a grant element of at least 25 percent. Values are presented in constant 2005 US dollars.

Official development assistance to agriculture

ODA to agriculture includes those commitments for the purposes of projects and programmes related to crops and livestock, forestry and fisheries. These include: *(crops and livestock)* agrarian reform, agricultural policy and administrative management, crop production, land and water resources, inputs, education, research, extension, training, plant and post-harvest protection and pest control, financial services, farmers' organizations and cooperatives, livestock production and veterinary services; *(forestry)* policy and administrative management, development, production of fuelwood and charcoal, education and training, research and services; *(fisheries)* policy and administrative management, development, education and training, research and services. The definition excludes rural development and development food aid.

Agricultural share of ODA to all sectors

ODA to agriculture divided by total ODA to all sectors multiplied by 100.

Unspecified recipients include all commitments made for which a recipient country or region was not specified.

Regional recipients represent the sum of all commitments of assistance to the following regions: Africa, America, Asia, Central Asia, Europe, Far East Asia, Middle East, North and Central America, Africa North of the Sahara, Oceania, South and Central Asia, South America, South Asia, Africa South of the Sahara and the West Indies.

Regional and income groupings

Countries are listed in alphabetical order according to the income and groupings established by the World Bank country classification system. A description of the World Bank country classifications is available at http://data.worldbank.org/about/country-classifications.

Country notes

Whenever possible, data from 1992 or 1995 onwards are shown for Armenia, Azerbaijan, Belarus, Estonia, Georgia, Kazakhstan, Kyrgyzstan, Latvia, Lithuania, Moldova, Russian Federation, Tajikistan, Turkmenistan, Ukraine, and Uzbekistan. Available data for years prior to 1992 are shown for the Union of Soviet Socialist Republics ("USSR" in the table listings).

Data for years prior to 1992 are provided for the former Yugoslavia ("Yugoslavia SFR" in the table listings). Observations for the years following 1992 are provided for the individual countries formed from the former Yugoslavia; these are Bosnia and Herzegovina, Croatia, the former Yugoslav Republic of Macedonia, and Slovenia, as well as Serbia and Montenegro. Observations are provided separately for Serbia and for Montenegro after the year 2006.

Data are shown when possible for the individual countries formed from the former Czechoslovakia – the Czech Republic and Slovakia. Data for years prior to 1993 are shown under Czechoslovakia.

Data are shown for Eritrea and Ethiopia separately, if possible; in most cases before 1992 data on Eritrea and Ethiopia are aggregated and presented as Ethiopia PDR.

Data for Yemen refer to that country from 1990 onward; data for previous years refer to aggregated data of the former People's Democratic Republic of Yemen and the former Yemen Arab Republic.

Separate observations are shown for Belgium and Luxembourg whenever possible.

Data for China exclude data for Hong Kong Special Administrative Region of China and Macao Special Administrative Region of China.

TABLE A1
Economically active population in agriculture and agricultural share of total economically active population, 1980, 1990, 2000 and 2010

	Economically active population in agriculture (*Thousands*)				Agricultural share of total economically active population (*Percentage*)			
	1980	1990	2000	2010	1980	1990	2000	2010
WORLD	**961 096**	**1 146 820**	**1 236 078**	**1 306 954**	**51**	**48**	**44**	**40**
LOW- AND MIDDLE-INCOME COUNTRIES	**920 209**	**1 114 313**	**1 212 473**	**1 289 537**	**61**	**58**	**53**	**47**
East Asia and the Pacific	**479 261**	**607 086**	**642 471**	**646 692**	**71**	**69**	**63**	**57**
American Samoa	5	7	8	8	45	41	36	29
Cambodia	2 337	3 138	4 028	4 966	76	74	70	66
China	380 386	482 507	504 849	500 977	74	72	67	61
Democratic People's Republic of Korea	3 136	3 618	3 328	3 065	44	38	30	23
Fiji	97	116	125	126	47	45	39	36
Indonesia	32 796	42 925	48 438	49 513	58	55	48	41
Kiribati	8	10	10	11	36	30	26	23
Lao People's Democratic Republic	1 166	1 486	1 865	2 368	80	78	77	75
Malaysia	2 048	1 933	1 849	1 612	41	27	19	13
Marshall Islands	..	..	6	6	..	..	25	23
Micronesia (Federated States of)	..	..	13	12	..	..	26	22
Mongolia	232	245	237	220	40	32	24	18
Myanmar	11 875	14 482	17 125	18 788	76	73	70	67
Palau	..	..	2	2	..	..	22	20
Papua New Guinea	1 063	1 421	1 725	2 110	83	80	75	69
Philippines	9 012	10 844	12 405	13 404	52	46	40	34
Samoa	26	24	22	18	48	43	35	27
Solomon Islands	66	90	118	151	78	75	72	68
Thailand	16 883	21 272	20 089	19 302	71	64	56	49
Timor-Leste	203	246	231	352	84	83	81	80
Tonga	12	12	12	11	50	41	33	27
Tuvalu	1	1	1	1	33	33	25	25
Vanuatu	26	30	33	38	49	43	37	30
Viet Nam	17 883	22 679	25 952	29 631	73	71	67	63
Europe and Central Asia	**45 311**	**42 919**	**32 580**	**27 449**	**26**	**23**	**18**	**14**
Albania	746	921	620	614	58	55	48	42
Armenia			174	148			13	9
Azerbaijan			972	1 085			27	23
Belarus			636	434			13	9
Bosnia and Herzegovina			100	44			5	2
Bulgaria	956	572	228	124	20	13	7	4
Georgia			472	354			20	15
Kazakhstan			1 321	1 192			17	14
Kyrgyzstan			543	510			26	21

TABLE A1 *(cont.)*

	Economically active population in agriculture *(Thousands)*				Agricultural share of total economically active population *(Percentage)*			
	1980	1990	2000	2010	1980	1990	2000	2010
Latvia			132	113			12	9
Lithuania			204	126			12	8
Montenegro				39				13
Republic of Moldova			390	200			23	15
Romania	3 680	2 603	1 739	868	35	24	15	9
Russian Federation			7 648	6 251			11	8
Serbia				617				13
Serbia and Montenegro			1 007				20	
Tajikistan			610	773			34	27
The former Yugoslav Republic of Macedonia			107	68			13	7
Turkey	8 205	10 355	9 131	8 067	56	51	41	32
Turkmenistan			627	705			33	30
Ukraine			3 295	2 412			14	10
USSR	29 983	27 557			22	19		
Uzbekistan			2 624	2 705			28	21
Yugoslav SFR	1 741	911			28	14		
Latin America and the Caribbean	**42 099**	**42 375**	**43 369**	**41 420**	**34**	**26**	**19**	**15**
Antigua and Barbuda	8	7	7	8	32	29	22	21
Argentina	1 309	1 458	1 458	1 405	13	12	9	7
Belize	15	18	25	31	38	33	27	24
Bolivia (Plurinational State of)	1 007	1 190	1 560	1 973	53	47	44	41
Brazil	16 342	14 062	13 325	11 049	37	24	16	11
Chile	764	934	962	964	20	19	16	13
Colombia	3 404	3 342	3 584	3 529	39	26	20	15
Costa Rica	274	307	326	322	32	26	20	15
Cuba	825	833	733	586	24	19	14	11
Dominica	9	8	7	6	33	30	24	21
Dominican Republic	567	621	547	457	32	25	16	10
Ecuador	984	1 117	1 210	1 228	39	32	24	19
El Salvador	632	655	661	590	40	32	28	23
Grenada	11	10	10	9	34	27	24	20
Guatemala	1 211	1 488	1 492	2 061	52	52	47	38
Guyana	67	58	55	50	26	22	17	15
Haiti	1 661	1 787	1 994	2 277	71	68	64	59
Honduras	649	672	735	665	57	41	31	24
Jamaica	296	275	248	214	31	25	21	17
Mexico	7 855	8 439	8 658	7 905	35	28	22	16
Nicaragua	382	391	390	351	38	29	21	15
Panama	191	247	258	248	29	27	20	16

TABLE A1 *(cont.)*

	Economically active population in agriculture *(Thousands)*				Agricultural share of total economically active population *(Percentage)*			
	1980	1990	2000	2010	1980	1990	2000	2010
Paraguay	493	576	715	831	39	34	29	25
Peru	2 185	2 773	3 344	3 692	39	33	29	24
Saint Lucia	13	15	16	17	34	28	23	20
Saint Vincent and the Grenadines	11	12	11	11	34	29	23	20
Suriname	25	29	30	33	24	21	19	17
Uruguay	191	184	197	186	15	14	13	11
Venezuela (Bolivarian Republic of)	718	867	811	722	15	13	8	5
Middle East and North Africa	**19 267**	**20 897**	**23 112**	**24 858**	**43**	**33**	**27**	**22**
Algeria	1 633	1 907	2 718	3 175	36	27	25	21
Djibouti	112	182	233	285	84	82	78	74
Egypt	6 411	6 495	6 339	6 620	54	40	31	25
Iran (Islamic Republic of)	4 260	5 040	5 761	6 553	39	32	27	22
Iraq	808	626	535	436	27	15	9	5
Jordan	76	102	118	114	16	14	9	6
Lebanon	121	69	48	28	14	7	4	2
Libya	188	127	103	71	22	10	6	3
Morocco	3 101	3 264	3 372	3 009	53	42	33	26
Occupied Palestinian Territory	111	128	125	110	24	18	12	8
Syrian Arab Republic	674	954	1 116	1 337	34	30	24	20
Tunisia	689	652	756	805	37	27	24	21
Yemen	1 083	1 351	1 888	2 315	68	56	48	39
South Asia	**228 463**	**269 218**	**307 395**	**348 834**	**68**	**63**	**58**	**53**
Afghanistan	3 258	2 804	4 485	6 046	70	68	64	60
Bangladesh	24 586	30 773	31 757	32 100	72	65	55	45
Bhutan	139	166	169	311	93	93	92	93
India	178 564	210 181	239 959	269 740	68	63	59	54
Maldives	24	20	21	23	52	34	23	15
Nepal	5 442	6 653	8 677	12 066	93	93	93	93
Pakistan	13 340	15 044	18 712	24 520	59	48	44	39
Sri Lanka	3 110	3 577	3 615	4 028	52	49	45	42
Sub-Saharan Africa	**105 808**	**131 818**	**163 546**	**200 284**	**72**	**68**	**63**	**58**
Angola	2 534	3 323	4 337	5 878	76	74	72	69
Benin	787	1 095	1 384	1 601	67	63	54	44
Botswana	206	206	281	317	61	45	44	42
Burkina Faso	2 894	3 742	4 982	6 909	92	92	92	92
Burundi	1 842	2 546	2 754	3 741	93	92	91	89
Cameroon	2 543	3 086	3 482	3 569	75	71	60	48
Cape Verde	35	34	35	32	37	30	23	17

TABLE A1 *(cont.)*

	Economically active population in agriculture *(Thousands)*				Agricultural share of total economically active population *(Percentage)*			
	1980	1990	2000	2010	1980	1990	2000	2010
Central African Republic	862	1 038	1 189	1 254	85	80	73	63
Chad	1 308	1 889	2 418	2 962	86	83	75	66
Comoros	104	135	171	222	80	78	73	69
Congo	397	447	501	524	57	48	40	32
Côte d'Ivoire	2 018	2 686	2 946	2 814	65	59	49	38
Democratic Republic of the Congo	7 504	9 460	11 694	14 194	71	67	62	57
Eritrea			1 090	1 547			77	74
Ethiopia			24 049	31 657			82	77
Ethiopia PDR	13 191	18 086			89	86		
Gabon	200	207	207	183	66	51	38	26
Gambia	236	351	461	605	85	82	79	76
Ghana	2 732	3 585	4 785	6 075	62	59	57	55
Guinea	1 913	2 372	3 320	3 832	91	87	84	80
Guinea-Bissau	289	338	391	447	88	85	82	79
Kenya	5 523	7 846	10 757	13 220	82	80	75	71
Lesotho	244	301	348	362	45	44	42	39
Liberia	550	568	712	913	77	72	67	62
Madagascar	3 196	4 029	5 243	7 255	82	79	75	70
Malawi	2 524	3 377	3 907	4 909	87	87	83	79
Mali	1 745	1 953	2 376	3 049	88	85	81	75
Mauritania	427	435	570	745	71	55	53	50
Mauritius	100	75	63	48	27	17	12	8
Mozambique	5 051	5 209	7 092	8 674	85	84	83	81
Namibia	177	219	253	267	57	50	41	34
Niger	1 756	2 247	3 099	4 237	90	88	86	83
Nigeria	12 790	12 689	12 443	12 267	54	43	33	25
Rwanda	2 156	2 824	3 242	4 360	93	92	91	89
Sao Tome and Principe	21	24	28	32	70	69	64	56
Senegal	1 839	2 296	2 929	3 821	80	76	74	70
Seychelles	23	25	28	30	85	81	80	75
Sierra Leone	894	1 083	1 041	1 326	73	71	65	60
Somalia	1 882	1 875	2 048	2 440	77	74	70	66
South Africa	1 606	1 614	1 482	1 188	17	13	9	6
Sudan	4 656	5 151	6 223	7 124	72	69	61	52
Swaziland	118	139	148	138	53	43	35	29
Togo	699	909	1 106	1 288	69	66	60	53
Uganda	4 946	6 665	8 420	11 016	87	85	80	75
United Republic of Tanzania	7 806	10 554	13 557	16 879	86	84	81	76
Zambia	1 483	2 215	2 685	3 215	75	74	69	63
Zimbabwe	2 001	2 870	3 269	3 118	73	69	63	56

TABLE A1 *(cont.)*

	Economically active population in agriculture *(Thousands)*				Agricultural share of total economically active population *(Percentage)*			
	1980	1990	2000	2010	1980	1990	2000	2010
HIGH-INCOME COUNTRIES	**40 855**	**32 470**	**23 567**	**17 379**	**10**	**7**	**5**	**3**
Andorra	3	3	2	2	19	13	7	5
Aruba	7	7	9	9	33	29	24	20
Australia	439	470	442	457	6	6	5	4
Austria	319	274	199	144	10	8	5	3
Bahamas	5	6	5	5	6	5	3	3
Bahrain	6	4	3	4	4	2	1	1
Barbados	11	9	7	4	10	7	5	2
Belgium			79	59			2	1
Belgium-Luxembourg	122	110			3	3		
Bermuda	1	1	1	1	4	3	3	3
Brunei Darussalam	4	2	1	1	6	2	1	1
Canada	806	495	382	332	7	3	2	2
Cayman Islands	2	3	4	5	33	30	24	21
China, Hong Kong SAR	..	..	..	..	..	..	..	..
China, Macao SAR	..	..	..	..	..	..	..	..
Croatia			170	84			8	4
Cyprus	81	50	38	30	26	14	9	5
Czech Republic			431	327			8	6
Czechoslovakia	1 077	985			13	12		
Denmark	184	162	108	75	7	6	4	3
Equatorial Guinea	67	108	142	176	77	73	69	64
Estonia			76	61			11	9
Faroe Islands	1	1	1	1	5	4	4	4
Finland	298	218	143	98	12	8	6	4
France	1 980	1 363	878	573	8	5	3	2
Germany	2 448	1 557	1 016	661	7	4	3	2
Greece	1 247	963	826	637	32	23	17	12
Greenland	1	1	1	0	4	3	3	0
Guam	16	20	19	20	37	32	26	23
Hungary	930	701	452	322	18	15	11	7
Iceland	12	15	13	12	10	11	8	6
Ireland	233	186	166	149	19	14	9	7
Israel	76	65	61	51	6	4	3	2
Italy	2 791	2 068	1 250	845	13	9	5	3
Japan	6 152	4 613	2 712	1 418	11	7	4	2
Kuwait	9	9	11	14	2	1	1	1
Liechtenstein	1	1	0	0	9	7	0	0
Luxembourg			4	3			2	1
Malta	10	3	3	2	8	2	2	1
Monaco	1	1	1	0	9	7	6	0

TABLE A1 *(cont.)*

	Economically active population in agriculture *(Thousands)*				Agricultural share of total economically active population *(Percentage)*			
	1980	1990	2000	2010	1980	1990	2000	2010
Netherlands	299	314	269	213	6	5	3	2
New Caledonia	24	30	32	32	49	43	36	30
New Zealand	150	171	175	186	11	10	9	8
Northern Mariana Islands	..	..	8	7	..	..	25	23
Norway	165	139	110	88	8	6	5	3
Oman	160	256	293	318	47	44	36	29
Poland	5 236	4 956	3 763	2 960	30	27	22	17
Portugal	1 170	857	678	515	26	18	13	9
Puerto Rico	54	49	30	16	6	4	2	1
Qatar	3	7	4	8	3	3	1	1
Republic of Korea	5 378	3 470	2 206	1 274	37	18	10	5
San Marino	2	1	1	1	22	9	8	7
Saudi Arabia	1 054	966	659	515	43	19	10	5
Singapore	17	6	3	2	2	0	0	0
Slovakia			240	197			9	7
Slovenia			19	7			2	1
Spain	2 626	1 890	1 339	1 015	18	12	7	4
Sweden	271	209	146	115	6	4	3	2
Switzerland	187	195	167	137	6	6	4	3
Trinidad and Tobago	46	50	50	47	11	11	9	7
Turks and Caicos Islands	1	1	2	3	33	25	25	18
United Arab Emirates	25	73	87	148	5	8	5	3
United Kingdom	715	639	529	475	3	2	2	1
United States of America	3 919	3 704	3 090	2 509	3	3	2	2
United States Virgin Islands	13	13	11	9	33	27	21	18

TABLE A2
Agricultural capital stock: total and per worker, 1980, 1990, 2000 and 2007

	Agricultural capital stock							
	Total (Million constant 2005 US$)				Per worker (Constant 2005 US$)			
	1980	1990	2000	2007	1980	1990	2000	2007
WORLD	**4 384 945**	**4 833 405**	**4 921 380**	**5 132 481**	**4 562**	**4 215**	**3 981**	**3 982**
LOW- AND MIDDLE-INCOME COUNTRIES	**2 654 288**	**3 014 823**	**3 143 266**	**3 365 730**	**2 884**	**2 706**	**2 592**	**2 610**
East Asia and the Pacific	**515 670**	**637 705**	**761 657**	**839 385**	**1 076**	**1 050**	**1 186**	**1 294**
American Samoa	12	12	12	12	2 340	1 686	1 516	1 518
Cambodia	1 969	4 239	4 942	5 439	842	1 351	1 227	1 149
China	347 912	420 169	499 079	540 792	915	871	989	1 071
Democratic People's Republic of Korea	5 712	7 091	6 743	7 065	1 821	1 960	2 026	2 236
Fiji	713	870	994	983	7 350	7 497	7 956	7 925
Indonesia	51 654	74 543	85 725	96 079	1 575	1 737	1 770	1 944
Kiribati	220	226	196	197	27 449	22 551	19 642	17 937
Lao People's Democratic Republic	1 975	2 627	3 164	3 746	1 694	1 768	1 696	1 705
Malaysia	13 563	18 595	20 661	21 095	6 623	9 620	11 174	12 453
Marshall Islands	..	..	38	38	..	..	6 332	6 337
Micronesia (Federated States of)	..	..	76	77	..	..	5 845	5 885
Mongolia	7 214	7 593	10 582	10 949	31 095	30 991	44 650	48 878
Myanmar	13 961	15 044	18 453	23 065	1 176	1 039	1 078	1 263
Palau	..	..	5	5	..	..	2 455	2 455
Papua New Guinea	1 720	1 895	2 294	2 385	1 618	1 334	1 330	1 206
Philippines	24 914	25 847	27 949	29 401	2 765	2 384	2 253	2 228
Samoa	381	313	302	333	14 668	13 043	13 746	17 544
Solomon Islands	148	166	176	192	2 235	1 841	1 491	1 368
Thailand	21 701	28 481	28 750	31 757	1 285	1 339	1 431	1 601
Timor-Leste	268	374	577	675	1 321	1 520	2 496	2 032
Tonga	136	117	118	123	11 360	9 710	9 873	10 217
Tuvalu	..	..	..	..	..	..	..	..
Vanuatu	422	496	566	631	16 226	16 517	17 155	17 066
Viet Nam	21 075	29 010	50 254	64 348	1 178	1 279	1 936	2 251
Europe and Central Asia	**727 033**	**762 671**	**583 169**	**559 847**	**16 045**	**17 770**	**17 900**	**19 433**
Albania	5 072	5 743	5 019	5 034	6 799	6 236	8 095	8 016
Armenia			2 657	2 879			15 267	18 575
Azerbaijan			12 419	12 984			12 776	12 284
Belarus			16 774	14 322			26 374	29 349
Bosnia and Herzegovina			1 892	1 931			18 919	33 879
Bulgaria	14 058	13 298	8 050	5 600	14 705	23 248	35 307	37 087
Georgia			6 056	5 410			12 831	14 200
Kazakhstan			43 093	46 002			32 622	37 800
Kyrgyzstan			6 260	6 216			11 529	12 117

TABLE A2 *(cont.)*

	Agricultural capital stock							
	Total (Million constant 2005 US$)				Per worker (Constant 2005 US$)			
	1980	1990	2000	2007	1980	1990	2000	2007
Latvia			3 164	3 538			23 969	29 733
Lithuania			7 624	7 899			37 372	55 237
Montenegro				390				8 666
Republic of Moldova			5 393	4 706			13 828	20 025
Romania	44 283	49 348	42 318	41 695	12 033	18 958	24 335	40 130
Russian Federation			185 689	161 586			24 279	24 280
Serbia				7 409				10 554
Serbia and Montenegro			8 251				8 193	
Tajikistan			5 700	6 295			9 345	8 553
The former Yugoslav Republic of Macedonia			1 297	1 448			12 124	18 328
Turkey	94 818	108 748	117 001	123 247	11 556	10 502	12 814	14 695
Turkmenistan			16 497	18 639			26 311	27 491
Ukraine			64 498	56 618			19 574	21 390
USSR	549 629	562 688			18 331	20 419		
Uzbekistan			23 518	25 997			8 963	9 639
Yugoslav SFR	19 174	22 846			11 013	25 077		
Latin America and the Caribbean	**581 207**	**635 421**	**667 946**	**710 649**	**13 806**	**14 995**	**15 401**	**16 761**
Antigua and Barbuda	34	35	36	38	4 189	4 943	5 104	4 746
Argentina	79 791	79 909	73 741	77 402	60 956	54 807	50 576	54 165
Belize	100	143	170	192	6 665	7 957	6 814	6 623
Bolivia (Plurinational State of)	6 459	6 126	7 606	9 122	6 414	5 148	4 876	4 931
Brazil	140 894	167 128	184 435	206 250	8 622	11 885	13 841	17 328
Chile	18 515	19 024	22 308	22 031	24 234	20 368	23 189	22 689
Colombia	88 886	93 958	97 034	101 981	26 112	28 114	27 074	28 582
Costa Rica	2 020	2 176	2 050	2 093	7 372	7 087	6 288	6 422
Cuba	27 877	27 827	24 985	23 913	33 791	33 406	34 086	38 017
Dominica	39	49	55	59	4 316	6 100	7 827	9 837
Dominican Republic	7 245	8 582	8 718	10 156	12 778	13 819	15 938	20 897
Ecuador	14 270	18 777	19 565	18 526	14 502	16 810	16 170	14 988
El Salvador	2 365	2 417	2 413	2 608	3 742	3 689	3 651	4 269
Grenada	43	35	34	38	3 876	3 488	3 386	4 246
Guatemala	5 358	5 727	6 738	9 203	4 425	3 849	4 516	4 710
Guyana	974	1 002	1 025	1 022	14 534	17 277	18 636	19 649
Haiti	3 938	3 688	4 813	4 887	2 371	2 064	2 414	2 218
Honduras	3 601	4 158	3 663	4 267	5 548	6 188	4 983	6 331
Jamaica	1 726	2 036	2 141	2 239	5 831	7 404	8 633	10 041
Mexico	100 140	111 384	117 366	118 762	12 749	13 199	13 556	14 501
Nicaragua	4 232	4 207	5 739	5 995	11 078	10 759	14 715	16 469
Panama	2 967	3 076	3 207	3 525	15 534	12 452	12 429	13 934

TABLE A2 *(cont.)*

	Agricultural capital stock							
	Total (*Million constant 2005 US$*)				Per worker (*Constant 2005 US$*)			
	1980	1990	2000	2007	1980	1990	2000	2007
Paraguay	4 676	6 599	7 536	8 318	9 485	11 457	10 540	10 411
Peru	19 148	19 548	22 071	23 350	8 763	7 049	6 600	6 447
Saint Lucia	54	65	66	56	4 140	4 327	4 126	3 292
Saint Vincent and the Grenadines	33	35	29	28	2 972	2 898	2 614	2 536
Suriname	567	712	750	662	22 668	24 549	24 986	20 678
Uruguay	24 426	21 436	22 124	24 972	127 885	116 499	112 306	132 829
Venezuela (Bolivarian Republic of)	20 830	25 564	27 528	28 955	29 011	29 486	33 944	38 351
Middle East and North Africa	**150 374**	**199 402**	**227 256**	**248 549**	**7 805**	**9 542**	**9 833**	**10 082**
Algeria	9 155	11 783	12 998	14 081	5 606	6 179	4 782	4 548
Djibouti	242	316	382	384	2 159	1 736	1 641	1 437
Egypt	22 484	25 714	32 377	35 992	3 507	3 959	5 108	5 429
Iran (Islamic Republic of)	46 137	67 144	74 309	82 643	10 830	13 322	12 899	12 841
Iraq	18 143	30 848	30 642	31 128	22 455	49 277	57 276	67 816
Jordan	793	1 155	1 388	1 492	10 434	11 324	11 759	13 086
Lebanon	2 488	2 601	2 749	2 774	20 560	37 695	57 264	84 063
Libya	4 612	7 005	6 945	7 309	24 534	55 158	67 426	90 229
Morocco	22 985	23 655	25 436	25 487	7 412	7 247	7 543	8 185
Occupied Palestinian Territory	364	421	640	676	3 279	3 293	5 122	5 925
Syrian Arab Republic	10 920	14 167	21 163	25 030	16 201	14 850	18 964	19 151
Tunisia	6 813	7 933	9 430	9 963	9 888	12 167	12 473	12 611
Yemen	5 239	6 661	8 797	11 594	4 838	4 930	4 659	5 241
South Asia	**399 171**	**460 007**	**531 857**	**583 962**	**1 747**	**1 709**	**1 730**	**1 733**
Afghanistan	26 818	27 213	30 437	30 398	8 232	9 705	6 786	5 397
Bangladesh	43 032	50 871	56 734	65 559	1 750	1 653	1 787	2 022
Bhutan	260	343	324	342	1 873	2 068	1 919	1 224
India	244 749	282 488	329 089	355 253	1 371	1 344	1 371	1 363
Maldives	12	15	18	15	485	726	839	636
Nepal	5 744	6 856	7 911	8 676	1 055	1 030	912	786
Pakistan	71 376	84 767	100 738	117 171	5 350	5 635	5 384	5 122
Sri Lanka	7 180	7 455	6 606	6 548	2 309	2 084	1 827	1 654
Sub-Saharan Africa	**280 833**	**319 616**	**371 382**	**423 337**	**2 654**	**2 425**	**2 271**	**2 248**
Angola	5 707	5 826	6 267	6 547	2 252	1 753	1 445	1 212
Benin	1 619	1 805	2 534	2 908	2 057	1 649	1 831	1 881
Botswana	1 841	2 100	2 045	1 845	8 937	10 192	7 279	6 151
Burkina Faso	3 357	5 222	7 596	10 079	1 160	1 396	1 525	1 610
Burundi	1 346	1 336	1 205	1 647	731	525	437	477
Cameroon	5 324	6 510	7 158	7 286	2 094	2 110	2 056	2 055
Cape Verde	76	123	153	206	2 169	3 605	4 364	6 246

TABLE A2 *(cont.)*

	Agricultural capital stock							
	Total (Million constant 2005 US$)				Per worker (Constant 2005 US$)			
	1980	1990	2000	2007	1980	1990	2000	2007
Central African Republic	1 269	1 693	2 171	2 460	1 472	1 631	1 826	2 012
Chad	4 267	4 329	5 667	7 033	3 262	2 292	2 344	2 468
Comoros	95	107	129	135	913	793	757	660
Congo	440	500	560	623	1 109	1 119	1 118	1 209
Côte d'Ivoire	4 435	6 392	7 108	7 563	2 198	2 380	2 413	2 669
Democratic Republic of the Congo	4 665	5 601	4 956	4 875	622	592	424	362
Eritrea			3 492	3 367			3 204	2 348
Ethiopia			32 771	48 465			1 363	1 649
Ethiopia PDR	29 785	33 338			2 258	1 843		
Gabon	375	410	452	429	1 875	1 982	2 182	2 235
Gambia	217	244	277	370	917	695	602	664
Ghana	3 876	4 431	5 748	7 025	1 419	1 236	1 201	1 249
Guinea	2 184	2 251	3 834	5 331	1 142	949	1 155	1 466
Guinea-Bissau	860	1 147	1 681	1 783	2 975	3 393	4 299	4 176
Kenya	12 632	17 295	15 958	18 301	2 287	2 204	1 484	1 463
Lesotho	986	1 036	1 132	1 070	4 042	3 441	3 253	2 998
Liberia	544	489	578	617	988	861	812	753
Madagascar	14 589	17 227	17 710	17 416	4 565	4 276	3 378	2 652
Malawi	1 596	1 870	2 462	3 066	633	554	630	680
Mali	7 067	6 658	9 891	12 499	4 050	3 409	4 163	4 395
Mauritania	2 558	2 914	3 969	4 306	5 990	6 699	6 963	6 240
Mauritius	225	247	260	267	2 248	3 294	4 134	5 046
Mozambique	2 838	3 580	4 405	4 843	562	687	621	592
Namibia	2 711	2 361	2 551	2 623	15 314	10 779	10 083	10 088
Niger	8 441	7 456	10 493	12 961	4 807	3 318	3 386	3 371
Nigeria	33 068	40 407	49 768	59 792	2 585	3 184	4 000	4 870
Rwanda	1 246	1 316	1 392	1 973	578	466	429	495
Sao Tome and Principe	191	201	216	218	9 099	8 368	7 711	7 042
Senegal	6 934	8 393	9 756	10 498	3 771	3 655	3 331	2 988
Seychelles	12	15	15	11	541	588	553	355
Sierra Leone	925	1 294	1 430	2 186	1 035	1 195	1 374	1 716
Somalia	11 621	13 440	13 088	13 145	6 175	7 168	6 391	5 663
South Africa	42 868	42 810	43 350	42 668	26 692	26 524	29 251	33 178
Sudan	24 999	27 681	43 260	47 540	5 369	5 374	6 952	7 002
Swaziland	801	876	852	809	6 788	6 299	5 754	5 824
Togo	907	1 452	1 549	1 747	1 298	1 597	1 400	1 417
Uganda	4 754	5 992	7 197	8 541	961	899	855	842
United Republic of Tanzania	15 058	16 679	19 829	21 504	1 929	1 580	1 463	1 372
Zambia	3 704	4 864	5 334	5 904	2 498	2 196	1 987	1 970
Zimbabwe	7 823	9 699	9 132	8 858	3 910	3 379	2 794	2 842

TABLE A2 *(cont.)*

	Agricultural capital stock							
	Total (*Million constant 2005 US$*)				Per worker (*Constant 2005 US$*)			
	1980	1990	2000	2007	1980	1990	2000	2007
HIGH-INCOME COUNTRIES	**1 730 513**	**1 818 454**	**1 776 270**	**1 764 612**	**42 328**	**55 944**	**75 328**	**92 456**
Andorra	0	0	0	0	147	147	220	220
Aruba	..	..	..	..	..	..	..	..
Australia	112 505	111 469	115 219	111 963	256 276	237 168	260 676	249 361
Austria	15 310	15 579	14 200	13 844	47 994	56 857	71 356	86 525
Bahamas	22	23	28	29	4 388	3 905	5 534	5 790
Bahrain	24	40	58	57	4 002	10 018	19 177	19 130
Barbados	102	127	108	84	9 307	14 079	15 366	16 892
Belgium			7 275	6 529			92 086	102 011
Belgium-Luxembourg	7 659	7 857			62 783	71 430		
Bermuda	..	..	..	..	..	..	..	..
Brunei Darussalam	45	37	58	81	11 190	18 425	58 130	81 360
Canada	88 391	91 794	91 090	94 170	109 666	185 442	238 455	271 384
Cayman Islands	..	..	..	..	..	..	..	..
China, Hong Kong SAR	..	..	..	..	..	..	..	..
China, Macao SAR	..	..	..	..	..	..	..	..
Croatia			..	..			..	..
Cyprus	814	954	1 092	1 141	10 052	19 087	28 733	34 582
Czech Republic			11 782	10 936			27 337	30 892
Czechoslovakia	20 886	23 151			19 393	23 503		
Denmark	16 591	14 061	12 292	11 906	90 170	86 796	113 816	141 738
Equatorial Guinea	408	410	413	355	6 084	3 799	2 908	2 152
Estonia			2 511	2 002			33 044	30 798
Faroe Islands	..	..	..	..	..	..	..	..
Finland	14 156	14 203	11 877	11 374	47 504	65 150	83 055	103 398
France	102 650	97 840	93 064	90 402	51 843	71 783	105 995	136 972
Germany	120 949	114 290	83 432	74 076	49 407	73 404	82 119	98 505
Greece	16 619	18 743	19 832	21 190	13 327	19 463	24 010	30 445
Greenland	5	6	5	5	5 140	5 690	5 450	5 310
Guam	28	28	28	29	1 776	1 423	1 498	1 434
Hungary	12 137	11 434	11 491	10 619	13 050	16 311	25 423	29 497
Iceland	1 006	954	809	936	83 863	63 614	62 239	77 962
Ireland	16 847	17 167	19 092	18 832	72 304	92 294	115 014	119 947
Israel	2 298	2 355	2 357	2 378	30 234	36 236	38 647	44 033
Italy	64 288	74 748	80 147	75 343	23 034	36 145	64 117	78 976
Japan	236 526	307 545	274 751	265 379	38 447	66 669	101 309	153 133
Kuwait	131	129	236	307	14 516	14 384	21 416	23 620
Liechtenstein	14	14	14	14	13 710	14 030	15 178	15 371
Luxembourg			505	440			126 143	146 760
Malta	62	95	90	96	6 153	31 537	30 113	48 245
Monaco	..	..	..	..	..	..	..	..

TABLE A2 *(cont.)*

	Agricultural capital stock							
	Total (Million constant 2005 US$)				Per worker (Constant 2005 US$)			
	1980	1990	2000	2007	1980	1990	2000	2007
Netherlands	13 026	13 442	12 382	11 816	43 565	42 808	46 031	51 376
New Caledonia	557	614	578	611	23 205	20 478	18 065	19 083
New Zealand	59 934	56 500	54 124	56 245	399 559	330 412	309 281	304 029
Northern Mariana Islands	..	..	..	..	..	..	..	..
Norway	8 227	9 076	8 467	8 270	49 858	65 297	76 975	88 924
Oman	673	946	1 264	1 311	4 208	3 694	4 313	4 444
Poland	50 722	65 865	65 784	71 100	9 687	13 290	17 482	22 323
Portugal	14 635	15 868	15 213	13 181	12 509	18 516	22 438	23 205
Puerto Rico	1 001	1 126	842	759	18 536	22 983	28 067	37 970
Qatar	56	108	210	189	18 560	15 397	52 535	27 049
Republic of Korea	6 085	9 355	14 238	16 248	1 132	2 696	6 454	10 739
San Marino	..	..	..	..	..	..	..	..
Saudi Arabia	9 053	21 277	23 127	23 239	8 589	22 026	35 093	39 590
Singapore	..	..	..	..	..	..	..	..
Slovakia			6 849	5 932			28 537	28 246
Slovenia			2 697	2 564			141 955	284 839
Spain	60 275	69 467	75 074	78 504	22 953	36 755	56 067	69 534
Sweden	15 582	14 089	13 835	13 394	57 496	67 412	94 760	108 896
Switzerland	8 770	8 877	8 113	7 983	46 898	45 524	48 580	54 678
Trinidad and Tobago	427	359	295	311	9 292	7 175	5 892	6 474
Turks and Caicos Islands	..	..	..	..	..	..	..	..
United Arab Emirates	769	1 031	3 309	3 670	30 766	14 128	38 040	30 085
United Kingdom	47 575	47 446	46 751	45 699	66 538	74 250	88 375	93 263
United States of America	582 673	557 953	569 262	579 069	148 679	150 635	184 227	216 799
United States Virgin Islands	21	18	17	16	1 650	1 352	1 525	1 649

TABLE A3

Average annual foreign direct investment inflows to agriculture, food, beverages and tobacco, and all sectors, 2005–06 and 2007–08

	Average annual foreign direct investment inflows (Million current US$)					
	To agriculture		To food, beverages and tobacco		To all sectors	
	2005–06	2007–08	2005–06	2007–08	2005–06	2007–08
WORLD						
LOW- AND MIDDLE-INCOME COUNTRIES						
East Asia and the Pacific						
American Samoa	..	..	..	..	..	..
Cambodia	72	95	18	20	432	841
China	659	886	..	2 611	70 937	83 582
Democratic People's Republic of Korea	..	..	..	..	..	..
Fiji	..	..	..	..	..	..
Indonesia	121	239	..	..	6 626	8 123
Kiribati	..	..	..	..	..	..
Lao People's Democratic Republic	7	..	..	..	28	..
Malaysia	–1	1 038	..	..	5 012	7 818
Marshall Islands	..	..	..	..	..	..
Micronesia (Federated States of)	..	..	..	..	..	..
Mongolia	..	..	..	..	..	..
Myanmar	..	..	..	..	71	..
Palau	..	..	..	..	..	..
Papua New Guinea	..	..	..	..	..	..
Philippines	0	2	..	..	2 388	2 160
Samoa	..	..	..	..	..	..
Solomon Islands	..	..	..	..	..	..
Thailand	5	10	46	194	8 536	8 923
Timor-Leste	..	..	..	..	..	..
Tonga	..	..	..	..	..	..
Tuvalu	..	..	..	..	..	..
Vanuatu	..	..	..	..	..	..
Viet Nam	56	..	..	..	2 021	..
Europe and Central Asia						
Albania	1	..	..	..	294	..
Armenia	..	23	19	23	370	900
Azerbaijan	..	..	..	..	3 911	4 144
Belarus	..	..	..	..	..	..
Bosnia and Herzegovina	..	..	59	69	690	1 570
Bulgaria	23	89	..	..	..	..
Georgia	..	..	..	..	..	..
Kazakhstan	1	..	24	..	1 971	..

TABLE A3 *(cont.)*

	Average annual foreign direct investment inflows (Million current US$)					
	To agriculture		To food, beverages and tobacco		To all sectors	
	2005–06	2007–08	2005–06	2007–08	2005–06	2007–08
Kyrgyzstan	–2	0	..	..	75	220
Latvia	1	51	..	..	1 185	1 792
Lithuania	11	13	–47	42	1 422	2 030
Montenegro		..		..		..
Republic of Moldova	1	4	..	..	162	249
Romania	56	159	307	196	8 923	11 916
Russian Federation	157	378	590	1 104	13 375	27 349
Serbia		40		147		3 466
Serbia and Montenegro	12		116		4 021	
Tajikistan	..	..	..	..	..	..
The former Yugoslav Republic of Macedonia	1	10	10	30	264	639
Turkey	7	25	338	1 009	13 087	16 935
Turkmenistan	..	..	..	..	..	..
Ukraine	..	..	..	..	..	..
USSR						
Uzbekistan	..	..	..	..	..	..
Yugoslav SFR						
Latin America and the Caribbean						
Antigua and Barbuda	..	..	..	..	..	..
Argentina	366	505	226	647	7 175	8 605
Belize	6	8	..	..	118	167
Bolivia (Plurinational State of)	0	3	..	..	535	1 020
Brazil	233	708	1 474	2 035	21 876	38 795
Chile	14	107	128	23	2 490	3 301
Colombia	7	41	..	..	8 454	9 816
Costa Rica	52	208	..	..	1 165	1 959
Cuba	..	..	..	..	..	..
Dominica	..	..	..	..	..	..
Dominican Republic	..	..	..	..	1 326	2 232
Ecuador	36	23	..	..	382	595
El Salvador	0	2	..	..	376	1 147
Grenada	..	..	..	..	..	..
Guatemala	..	..	53	88	550	724
Guyana	..	..	..	..	..	..
Haiti	..	..	..	..	..	..
Honduras	48	7	..	..	537	903
Jamaica	0	..	..	..	782	1 152
Mexico	16	82	2 175	1 344	20 789	24 806

TABLE A3 *(cont.)*

	Average annual foreign direct investment inflows *(Million current US$)*					
	To agriculture		To food, beverages and tobacco		To all sectors	
	2005–06	2007–08	2005–06	2007–08	2005–06	2007–08
Nicaragua	8	..	..	..	264	504
Panama	..	..	..	..	1 737	1 777
Paraguay	–18	1	13	–4	75	157
Peru	62	30	..	..	723	1 234
Saint Lucia	..	..	..	..	..	..
Saint Vincent and the Grenadines	..	..	..	..	..	..
Suriname	..	..	..	..	..	..
Uruguay	283	335	11	100	1 170	1 330
Venezuela (Bolivarian Republic of)	..	..	..	..	1 000	646
Middle East and North Africa						
Algeria	..	..	..	..	..	..
Djibouti	..	..	..	..	..	..
Egypt	30	100	..	..	13 084	15 319
Iran (Islamic Republic of)	..	..	..	..	..	..
Iraq	..	..	..	..	..	..
Jordan	..	..	..	..	..	..
Lebanon	..	..	..	..	..	..
Libya	..	..	..	..	..	..
Morocco	2	5	..	..	2 988	4 121
Occupied Palestinian Territory	..	..	..	..	..	..
Syrian Arab Republic	6	15	..	..	621	1 355
Tunisia	8	11	12	22	2 045	2 187
Yemen	..	..	..	..	..	..
South Asia						
Afghanistan	..	..	..	..	..	..
Bangladesh	2	11	5	16	819	876
Bhutan	..	..	..	..	..	..
India	..	..	..	..	6 333	21 062
Maldives	..	..	..	..	..	..
Nepal	..	..	..	..	..	..
Pakistan	..	..	56	298	3 236	5 514
Sri Lanka	..	..	..	..	..	..
Sub-Saharan Africa						
Angola	..	..	..	..	..	..
Benin	..	..	..	..	..	..
Botswana	..	..	..	..	..	..
Burkina Faso	..	..	..	..	..	..

TABLE A3 *(cont.)*

	Average annual foreign direct investment inflows (Million current US$)					
	To agriculture		To food, beverages and tobacco		To all sectors	
	2005–06	2007–08	2005–06	2007–08	2005–06	2007–08
Burundi	..	..	..	..	..	..
Cameroon	..	..	..	..	..	..
Cape Verde	..	..	..	..	..	..
Central African Republic	..	..	..	..	..	..
Chad	..	..	..	..	..	..
Comoros	..	..	..	..	..	..
Congo	..	..	..	..	..	..
Côte d'Ivoire	..	..	..	..	..	..
Democratic Republic of the Congo	..	..	..	..	..	..
Eritrea	..	..	..	..	..	..
Ethiopia	..	..	..	..	..	..
Ethiopia PDR						
Gabon	..	..	..	..	..	..
Gambia	..	..	..	..	..	..
Ghana	..	..	..	..	..	..
Guinea	..	..	..	..	..	..
Guinea-Bissau	..	..	..	..	..	..
Kenya	..	..	..	..	..	..
Lesotho	..	..	..	..	..	..
Liberia	..	..	..	..	..	..
Madagascar	8	–6	..	..	190	979
Malawi	..	..	..	..	..	..
Mali	..	..	..	..	..	..
Mauritania	..	..	..	..	509	..
Mauritius	1	8	..	..	162	385
Mozambique	9	71	..	..	131	510
Namibia	..	..	..	..	..	..
Niger	..	..	..	..	..	..
Nigeria	..	..	..	..	3 403	..
Rwanda	..	..	..	..	..	..
Sao Tome and Principe	..	..	..	..	..	..
Senegal	..	..	..	..	..	..
Seychelles	..	..	..	..	..	..
Sierra Leone	..	..	..	..	..	..
Somalia	..	..	..	..	..	..
South Africa	..	..	..	..	..	..
Sudan	..	..	..	..	..	..
Swaziland	..	..	..	..	..	..
Togo	..	..	..	..	..	..
Uganda	..	..	..	..	..	..

TABLE A3 *(cont.)*

	Average annual foreign direct investment inflows (Million current US$)					
	To agriculture		To food, beverages and tobacco		To all sectors	
	2005–06	2007–08	2005–06	2007–08	2005–06	2007–08
United Republic of Tanzania	11	..	..	..	448	..
Zambia	..	..	..	..	..	..
Zimbabwe	..	..	..	..	..	..
HIGH-INCOME COUNTRIES						
Andorra	..	..	..	..	..	..
Aruba	..	..	..	..	..	..
Australia	–107	–9	..	..	–3 109	34 207
Austria	–20	4	290	–511	9 634	19 006
Bahamas	..	..	..	..	..	..
Bahrain	..	..	..	..	..	..
Barbados	..	..	..	..	..	..
Belgium	–973	–92	..	..	34 373	110 099
Belgium-Luxembourg						
Brunei Darussalam	0	..	..	..	289	248
Canada	..	..	..	..	42 993	84 961
Cayman Islands	..	..	..	..	..	..
China, Hong Kong SAR	..	..	..	..	39 341	56 981
China, Macao SAR	..	..	..	..	1 424	2 448
Croatia	11	4	120	101	2 654	5 581
Cyprus	0	..	2	0	1 525	3 142
Czech Republic	32	0	138	392	8 558	8 447
Czechoslovakia						
Denmark	0	..	–8	2 763	7 775	7 261
Equatorial Guinea	..	..	..	..	..	..
Estonia	18	20	..	..	2 333	2 331
Faroe Islands	..	..	..	..	..	..
Finland	..	..	..	..	6 201	5 205
France	44	33	5 281	3 392	78 397	79 230
French Polynesia	..	..	..	..	..	..
Germany	11	8	732	–639	51 533	51 514
Greece	34	4	28	–109	2 989	3 305
Greenland	..	..	..	..	..	..
Guam	..	..	..	..	..	..
Hungary	8	32	80	–106	7 263	5 668
Iceland	0	–2	127	24	3 550	3 557
Ireland	..	..	–66	–1 797	–18 616	2 339
Israel	..	..	23	71	9 303	9 665
Italy	–74	149	2 114	–244	24 336	30 863
Japan	–15	4	–474	94	–1 865	23 487

TABLE A3 *(cont.)*

	Average annual foreign direct investment inflows *(Million current US$)*					
	To agriculture		To food, beverages and tobacco		To all sectors	
	2005–06	2007–08	2005–06	2007–08	2005–06	2007–08
Kuwait	..	..	..	..	..	..
Liechtenstein	..	..	..	..	..	..
Luxembourg	..	..	..	..	..	..
Malta	..	..	..	..	1 239	885
Monaco	..	..	..	..	..	..
Netherlands	..	..	–338	10 392	27 622	55 742
New Caledonia	..	..	..	..	..	..
New Zealand	..	..	..	..	..	..
Northern Mariana Islands	..	..	..	..	..	..
Norway	..	..	..	..	4 426	4 893
Oman	..	..	..	..	1 746	3 200
Poland	52	117	499	416	14 906	22 695
Portugal	..	..	..	..	7 419	3 864
Puerto Rico	..	..	..	..	..	..
Qatar	..	..	..	..	..	..
Republic of Korea	2	..	–150	..	6 000	..
San Marino	..	..	..	..	..	..
Saudi Arabia	8	24	–542	179	15 195	31 270
Singapore	..	..	34	50	2 183	–479
Slovakia	2	1	..	..	2 703	3 267
Slovenia	..	..	..	..	..	..
Spain	–4	..	..	..	22 518	0
Sweden	..	..	24	4 435	20 418	32 114
Switzerland	..	..	..	..	21 383	23 792
Trinidad and Tobago	..	..	6	10	911	830
Turks and Caicos Islands	..	..	..	..	..	..
United Arab Emirates	..	..	..	..	..	..
United Kingdom	88	79	1 959	10 468	166 096	136 618
United States of America	22	240	8 619	29 025	170 955	293 644
United States Virgin Islands	..	..	..	..	..	..

TABLE A4

Government expenditures: total spent on agriculture and agricultural share of total expenditures, 1980, 1990, 2000 and 2007

	Government expenditures							
	Total spent on agriculture *(Million constant 2005 PPP dollars)*				Agricultural share of total expenditures *(Percentage)*			
	1980	1990	2000	2007	1980	1990	2000	2007
LOW- AND MIDDLE-INCOME COUNTRIES								
East Asia and the Pacific								
American Samoa	..	..	..	..	..	..	..	..
Cambodia	..	..	..	..	..	..	..	..
China	16 618	20 567	41 743	88 683	12.2	10.0	7.8	6.8
Democratic People's Republic of Korea	..	..	..	..	..	..	..	..
Fiji	27	34	36	31	7.2	6.7	3.7	3.1
Indonesia	4 061	4 851	2 671	3 856	10.0	7.6	2.3	3.0
Kiribati	..	..	..	..	..	..	..	..
Lao People's Democratic Republic	..	..	..	..	..	..	..	..
Malaysia	..	..	..	..	..	..	..	..
Marshall Islands	..	..	..	..	..	..	..	..
Micronesia (Federated States of)	..	..	..	..	..	..	..	..
Mongolia	..	..	..	..	..	..	..	..
Myanmar	403	183	391	420	23.6	9.3	17.4	8.3
Palau	..	..	..	..	..	..	..	..
Papua New Guinea	141	152	78	50	8.5	7.2	2.6	1.5
Philippines	1 020	1 986	2 217	2 550	6.1	6.6	5.7	5.2
Samoa	..	..	..	..	..	..	..	..
Solomon Islands	..	..	..	..	..	..	..	..
Thailand	1 917	3 301	5 510	6 311	9.7	10.4	8.8	6.6
Timor-Leste	..	..	..	..	..	..	..	..
Tonga	..	..	..	..	..	..	..	..
Tuvalu	..	..	..	..	..	..	..	..
Vanuatu	4	9	6	6	3.0	4.6	3.6	5.0
Viet Nam	..	..	..	..	..	..	..	..
Europe and Central Asia								
Albania	..	..	..	..	..	..	..	..
Armenia			..	..			..	..
Azerbaijan			..	..			..	..
Belarus			1 397	2 840			13.1	10.5
Bosnia and Herzegovina			..	..			..	..
Bulgaria	..	..	..	..	..	..	..	..
Georgia			..	..			..	..
Kazakhstan			236	1 040			2.1	4.1
Kyrgyzstan			54	58			4.0	2.4
Latvia			269	1 071			4.5	9.9
Lithuania			133	701			2.8	3.9

TABLE A4 *(cont.)*

	Government expenditures							
	Total spent on agriculture *(Million constant 2005 PPP dollars)*				**Agricultural share of total expenditures** *(Percentage)*			
	1980	**1990**	**2000**	**2007**	**1980**	**1990**	**2000**	**2007**
Montenegro				..				..
Republic of Moldova			55	198			4.3	8.1
Romania	..	..	1 763	3 248	..	..	3.4	4.9
Russian Federation			3 763	1 881			2.1	0.5
Serbia				..				..
Serbia and Montenegro			..				..	
Tajikistan			..	..			..	..
The former Yugoslav Republic of Macedonia			..	..			..	..
Turkey	..	..	4 557	4 595	..	..	2.8	2.2
Turkmenistan			..	..			..	..
Ukraine			..	..			..	..
USSR	..	..			..	..		
Uzbekistan			..	..			..	..
Yugoslav SFR	..	..			..	..		
Latin America and the Caribbean								
Antigua and Barbuda	..	..	..	..	..	..	..	..
Argentina	..	..	..	..	..	..	..	..
Belize	..	..	..	..	..	..	..	..
Bolivia (Plurinational State of)	80	74	150	109	3.4	3.5	2.8	1.4
Brazil	10	0	3 497	2 386	6.6	1.5	4.9	2.1
Chile	..	..	..	..	..	..	..	..
Colombia	..	..	..	..	..	..	..	..
Costa Rica	121	193	132	228	3.4	4.1	3.4	3.5
Cuba	..	..	..	..	..	..	..	..
Dominica	..	..	..	..	..	..	..	..
Dominican Republic	382	343	475	220	14.3	14.5	7.6	3.4
Ecuador	..	..	..	..	..	..	..	..
El Salvador	478	952	1 637	87	5.8	5.4	5.6	1.5
Grenada	..	..	..	..	..	..	..	..
Guatemala	251	145	145	204	6.6	4.2	2.3	2.4
Guyana	..	..	..	..	..	..	..	..
Haiti	..	..	..	..	..	..	..	..
Honduras	..	..	..	..	..	..	..	..
Jamaica	..	..	..	..	..	..	..	..
Mexico	7 951	4 579	4 712	6 794	7.2	3.5	3.3	1.8
Nicaragua	..	..	..	..	..	..	..	..
Panama	180	76	80	206	5.3	2.5	1.5	2.8
Paraguay	..	..	..	..	..	..	..	..
Peru	..	..	..	..	..	..	..	..

TABLE A4 *(cont.)*

	Government expenditures							
	Total spent on agriculture *(Million constant 2005 PPP dollars)*				Agricultural share of total expenditures *(Percentage)*			
	1980	1990	2000	2007	1980	1990	2000	2007
Saint Lucia	..	..	..	..	..	..	..	..
Saint Vincent and the Grenadines	3	12	8	6	3.8	6.3	3.2	2.3
Suriname	..	..	..	..	..	..	..	..
Uruguay	57	64	106	113	2.1	1.3	1.2	1.2
Venezuela (Bolivarian Republic of)	..	..	..	..	..	..	..	..
Middle East and North Africa								
Algeria	..	..	..	..	..	..	..	..
Djibouti	..	..	..	..	..	..	..	..
Egypt	2 522	2 387	4 843	3 122	4.6	5.4	6.8	3.0
Iran (Islamic Republic of)	2 713	2 324	1 947	5 985	3.4	3.4	1.9	3.1
Iraq	..	..	..	..	..	..	..	..
Jordan	37	90	175	154	1.0	2.4	3.2	1.5
Lebanon	..	..	..	..	..	..	..	..
Libya	..	..	..	..	..	..	..	..
Morocco	880	863	833	725	6.5	5.3	3.2	2.0
Occupied Palestinian Territory	..	..	..	..	..	..	..	..
Syrian Arab Republic	751	945	1 742	1 338	5.0	11.0	9.5	5.6
Tunisia	1 061	910	1 280	1 076	14.5	9.6	9.3	6.0
Yemen	9	111	206	201	1.4	2.2	1.7	1.0
South Asia								
Afghanistan	..	..	..	..	..	..	..	..
Bangladesh	392	498	528	1 229	10.3	4.7	3.6	5.6
Bhutan	32	48	53	59	31.9	14.5	8.0	5.0
India	5 415	14 058	15 695	23 457	7.2	8.3	5.6	5.0
Maldives	5	3	6	20	8.8	1.9	1.7	2.2
Nepal	219	183	188	191	16.4	8.5	5.8	4.3
Pakistan	308	278	356	2 950	2.1	0.8	0.7	4.1
Sri Lanka	500	546	611	842	5.8	5.8	4.3	4.4
Sub-Saharan Africa								
Angola	..	..	..	..	..	..	..	..
Benin	..	..	..	..	..	..	..	..
Botswana	100	196	281	221	9.7	6.5	4.2	2.7
Burkina Faso	..	..	..	..	..	..	..	..
Burundi	..	..	..	..	..	..	..	..
Cameroon	..	..	..	..	..	..	..	..
Cape Verde	..	..	..	..	..	..	..	..
Central African Republic	..	..	..	..	..	..	..	..
Chad	..	..	..	..	..	..	..	..

TABLE A4 *(cont.)*

	Government expenditures							
	Total spent on agriculture *(Million constant 2005 PPP dollars)*				Agricultural share of total expenditures *(Percentage)*			
	1980	1990	2000	2007	1980	1990	2000	2007
Comoros	..	..	..	..	..	..	..	..
Congo	..	..	..	..	..	..	..	..
Côte d'Ivoire	..	..	..	..	..	..	..	..
Democratic Republic of the Congo	..	..	..	..	..	..	..	..
Eritrea			..	..			..	..
Ethiopia			..	..			..	..
Ethiopia PDR	..	..			..	..		
Gabon	..	..	..	..	..	..	..	..
Gambia	..	..	..	..	..	..	..	..
Ghana	139	6	39	54	12.2	0.4	0.7	0.4
Guinea	..	..	..	..	..	..	..	..
Guinea-Bissau	..	..	..	..	..	..	..	..
Kenya	331	652	390	425	8.3	10.2	5.5	3.4
Lesotho	27	82	44	49	8.0	9.8	3.7	3.2
Liberia	..	..	..	..	..	..	..	..
Madagascar	..	..	..	..	..	..	..	..
Malawi	149	153	101	83	10.2	9.9	4.9	4.1
Mali	..	..	..	..	..	..	..	..
Mauritania	..	..	..	..	..	..	..	..
Mauritius	68	101	116	84	6.9	7.3	4.8	2.7
Mozambique	..	..	..	..	..	..	..	..
Namibia	..	..	..	..	..	..	..	..
Niger	..	..	..	..	..	..	..	..
Nigeria	936	796	419	510	3.0	5.1	2.0	2.0
Rwanda	..	..	..	..	..	..	..	..
Sao Tome and Principe	..	..	..	..	..	..	..	..
Senegal	..	..	..	..	..	..	..	..
Seychelles	..	..	..	..	..	..	..	..
Sierra Leone	..	..	..	..	..	..	..	..
Somalia	..	..	..	..	..	..	..	..
South Africa	..	..	..	..	..	..	..	..
Sudan	..	..	..	..	..	..	..	..
Swaziland	30	47	80	121	13.0	7.3	6.6	4.4
Togo	..	..	..	..	..	..	..	..
Uganda	27	42	223	231	6.7	2.3	6.3	4.0
United Republic of Tanzania	..	..	..	..	..	..	..	..
Zambia	832	83	216	333	22.9	2.8	6.5	8.3
Zimbabwe	..	..	..	..	..	..	..	..

TABLE A5
Government expenditures on agriculture: per agricultural worker and Agricultural Orientation Index, 1980, 1990, 2000 and 2007

	Government expenditures							
	Per agricultural worker *(Constant 2005 PPP dollars)*				Agricultural Orientation Index *(Ratio)*			
	1980	1990	2000	2007	1980	1990	2000	2007
LOW- AND MIDDLE-INCOME COUNTRIES								
East Asia and the Pacific								
American Samoa	..	..	..	..	..	..	..	..
Cambodia	..	..	..	..	..	..	..	..
China	45	43	84	178	0.40	0.37	0.51	0.64
Democratic People's Republic of Korea	..	..	..	..	..	..	..	..
Fiji	283	294	288	248	0.33	0.33	0.22	0.22
Indonesia	127	117	57	81	0.42	0.39	0.15	0.21
Kiribati	..	..	..	..	..	..	..	..
Lao People's Democratic Republic	..	..	..	..	..	..	..	..
Malaysia	..	..	..	..	..	..	..	..
Marshall Islands	..	..	..	..	..	..	..	..
Micronesia (Federated States of)	..	..	..	..	..	..	..	..
Mongolia	..	..	..	..	..	..	..	..
Myanmar	33	12	22	22	..	..	..	..
Palau	..	..	..	..	..	..	..	..
Papua New Guinea	133	108	45	25	0.24	0.23	0.07	0.04
Philippines	111	181	178	193	0.24	0.30	0.36	0.37
Samoa	..	..	..	..	..	..	..	..
Solomon Islands	..	..	..	..	..	..	..	..
Thailand	114	156	278	322	0.42	0.84	0.98	0.62
Timor-Leste	..	..	..	..	..	..	..	..
Tonga	..	..	..	..	..	..	..	..
Tuvalu	..	..	..	..	..	..	..	..
Vanuatu	130	302	163	157	0.16	0.22	0.14	0.22
Viet Nam	..	..	..	..	..	..	..	..
Europe and Central Asia								
Albania	..	..	..	..	..	..	..	..
Armenia			..	..			..	..
Azerbaijan			..	..			..	..
Belarus			2 200	5 819			0.92	1.13
Bosnia and Herzegovina			..	..			..	..
Bulgaria	..	..	..	..	..	..	..	..
Georgia			..	..			..	..
Kazakhstan			179	860			0.24	0.67
Kyrgyzstan			99	108			0.11	0.08
Latvia			2 040	9 079			0.99	2.76
Lithuania			650	4 934			0.45	1.01

TABLE A5 *(cont.)*

	Government expenditures							
	Per agricultural worker *(Constant 2005 PPP dollars)*				Agricultural Orientation Index *(Ratio)*			
	1980	1990	2000	2007	1980	1990	2000	2007
Montenegro				..				..
Republic of Moldova			142	842			0.15	0.67
Romania	..	..	1 016	3 153	..	..	0.27	0.56
Russian Federation			492	285			0.32	0.11
Serbia				..				..
Serbia and Montenegro			..				..	
Tajikistan			..	..			..	..
The former Yugoslav Republic of Macedonia			..	..			..	..
Turkey	..	..	478	525	..	..	0.25	0.26
Turkmenistan			..	..			..	..
Ukraine			..	..			..	..
USSR	..	..			..	..		
Uzbekistan			..	..			..	..
Yugoslav SFR	..	..			..	..		
Latin America and the Caribbean								
Antigua and Barbuda	..	..	..	..	..	..	..	..
Argentina	..	..	..	..	..	..	..	..
Belize	..	..	..	..	..	..	..	..
Bolivia (Plurinational State of)	79	62	96	59	0.18	0.21	0.19	0.11
Brazil	1	0	263	200	0.60	0.19	0.87	0.38
Chile	..	..	..	..	..	..	..	..
Colombia	..	..	..	..	..	..	..	..
Costa Rica	442	628	403	698	..	..	..	..
Cuba	..	..	..	..	..	..	..	..
Dominica	..	..	..	..	..	..	..	..
Dominican Republic	658	539	844	440	0.71	1.08	1.05	0.52
Ecuador	..	..	..	..	..	..	..	..
El Salvador	755	1 454	2 480	142	..	..	..	..
Grenada	..	..	..	..	..	..	..	..
Guatemala	208	98	97	105	..	..	..	..
Guyana	..	..	..	..	..	..	..	..
Haiti	..	..	..	..	..	..	..	..
Honduras	..	..	..	..	..	..	..	..
Jamaica	..	..	..	..	..	..	..	..
Mexico	1 011	549	547	843	0.79	0.45	0.80	0.49
Nicaragua	..	..	..	..	..	..	..	..
Panama	942	309	311	816	0.59	0.26	0.21	0.42
Paraguay	..	..	..	..	..	..	..	..
Peru	..	..	..	..	..	..	..	..

TABLE A5 *(cont.)*

	Government expenditures							
	Per agricultural worker *(Constant 2005 PPP dollars)*				Agricultural Orientation Index *(Ratio)*			
	1980	1990	2000	2007	1980	1990	2000	2007
Saint Lucia	..	..	..	..	..	..	..	..
Saint Vincent and the Grenadines	309	984	765	561	0.27	0.30	0.29	0.29
Suriname	..	..	..	..	..	..	..	..
Uruguay	296	347	536	600	..	..	..	..
Venezuela (Bolivarian Republic of)	..	..	..	..	..	..	..	..
Middle East and North Africa								
Algeria	..	..	..	..	..	..	..	..
Djibouti	..	..	..	..	..	..	..	..
Egypt	398	361	736	452	0.25	0.28	0.41	0.22
Iran (Islamic Republic of)	629	442	329	917	0.20	0.18	0.14	0.30
Iraq	..	..	..	..	..	..	..	..
Jordan	497	927	1 467	1 283	0.12	0.29	1.38	0.54
Lebanon	..	..	..	..	..	..	..	..
Libya	..	..	..	..	..	..	..	..
Morocco	284	264	247	231	0.35	0.29	0.21	0.15
Occupied Palestinian Territory	..	..	..	..	..	..	..	..
Syrian Arab Republic	1 106	959	1 511	964	..	..	..	..
Tunisia	1 538	1 394	1 691	1 367	1.03	0.61	0.75	0.59
Yemen	8	79	106	90	..	..	..	..
South Asia								
Afghanistan	..	..	..	..	..	..	..	..
Bangladesh	14	15	15	35	0.33	0.15	0.14	0.29
Bhutan	232	293	320	214	..	..	..	..
India	31	68	66	91	0.20	0.28	0.24	0.28
Maldives	217	183	292	865	..	..	..	..
Nepal	40	27	22	17	0.27	0.16	0.14	0.13
Pakistan	22	18	18	122	0.07	0.03	0.03	0.20
Sri Lanka	161	153	169	217	0.21	0.22	0.22	0.38
Sub-Saharan Africa								
Angola	..	..	..	..	..	..	..	..
Benin	..	..	..	..	..	..	..	..
Botswana	490	973	1 024	750	0.66	1.33	1.55	1.32
Burkina Faso	..	..	..	..	..	..	..	..
Burundi	..	..	..	..	..	..	..	..
Cameroon	..	..	..	..	..	..	..	..
Cape Verde	..	..	..	..	..	..	..	..
Central African Republic	..	..	..	..	..	..	..	..
Chad	..	..	..	..	..	..	..	..

TABLE A5 *(cont.)*

	Government expenditures							
	Per agricultural worker *(Constant 2005 PPP dollars)*				Agricultural Orientation Index *(Ratio)*			
	1980	**1990**	**2000**	**2007**	**1980**	**1990**	**2000**	**2007**
Comoros	..	..	..	..	..	..	..	..
Congo	..	..	..	..	..	..	..	..
Côte d'Ivoire	..	..	..	..	..	..	..	..
Democratic Republic of the Congo	..	..	..	..	..	..	..	..
Eritrea			..	..			..	..
Ethiopia			..	..			..	..
Ethiopia PDR	..	..			..	..		
Gabon	..	..	..	..	..	..	..	..
Gambia	..	..	..	..	..	..	..	..
Ghana	51	2	8	10	0.20	0.01	0.02	0.01
Guinea	..	..	..	..	..	..	..	..
Guinea-Bissau	..	..	..	..	..	..	..	..
Kenya	60	83	36	34	0.25	0.35	0.17	0.17
Lesotho	110	277	129	142	0.33	0.39	0.30	0.39
Liberia	..	..	..	..	..	..	..	..
Madagascar	..	..	..	..	..	..	..	..
Malawi	59	45	25	17	0.23	0.22	0.13	0.12
Mali	..	..	..	..	..	..	..	..
Mauritania	..	..	..	..	..	..	..	..
Mauritius	672	1 351	1 845	1 580	0.52	0.57	0.69	0.56
Mozambique	..	..	..	..	..	..	..	..
Namibia	..	..	..	..	..	..	..	..
Niger	..	..	..	..	..	..	..	..
Nigeria	74	63	33	41	..	..	..	..
Rwanda	..	..	..	..	..	..	..	..
Sao Tome and Principe	..	..	..	..	..	..	..	..
Senegal	..	..	..	..	..	..	..	..
Seychelles	..	..	..	..	..	..	..	..
Sierra Leone	..	..	..	..	..	..	..	..
Somalia	..	..	..	..	..	..	..	..
South Africa	..	..	..	..	..	..	..	..
Sudan	..	..	..	..	..	..	..	..
Swaziland	253	341	531	857	0.57	0.70	0.53	0.60
Togo	..	..	..	..	..	..	..	..
Uganda	6	6	26	23	0.09	0.04	0.21	0.17
United Republic of Tanzania	..	..	..	..	..	..	..	..
Zambia	561	37	78	109	1.52	0.14	0.29	0.39
Zimbabwe	..	..	..	..	..	..	..	..

TABLE A6

Public expenditures on agricultural research and development: total and as a share of agricultural GDP, 1981, 1990, 2000 and latest year

	Public expenditures on agricultural research and development							
	Total (*Million constant 2005 PPP dollars*)				As a share of agricultural GDP (*Percentage*)			
	1981	**1990**	**2000**	**Latest year**	**1981**	**1990**	**2000**	**Latest year**
LOW- AND MIDDLE-INCOME COUNTRIES								
East Asia and Pacific								
American Samoa	..	..	..	..	..	..	..	..
Cambodia	..	..	..	..	..	..	..	..
China	658	1 055	1 745	4 048	0.41	0.34	0.38	0.50
Democratic People's Republic of Korea	..	..	..	..	..	..	..	..
Fiji	..	..	..	..	..	..	..	..
Indonesia	..	..	154	204	..	..	0.18	0.20
Kiribati	..	..	..	..	..	..	..	..
Lao People's Democratic Republic	..	..	22	10	..	..	0.57	0.24
Malaysia	158	210	335	..	1.01	1.14	1.57	..
Marshall Islands	..	..	..	..	..	..	..	..
Micronesia (Federated States of)	..	..	..	..	..	..	..	..
Mongolia	..	..	..	..	..	..	..	..
Myanmar	..	..	6	5	..	..	0.04	0.06
Palau	..	..	..	..	..	..	..	..
Papua New Guinea	..	..	21	..	..	..	0.60	..
Philippines	..	..	129	..	..	..	0.41	..
Samoa	..	..	..	..	..	..	..	..
Solomon Islands	..	..	..	..	..	..	..	..
Thailand	..	..	..	..	..	..	..	..
Timor-Leste	..	..	..	..	..	..	..	..
Tonga	..	..	..	..	..	..	..	..
Tuvalu	..	..	..	..	..	..	..	..
Vanuatu	..	..	..	..	..	..	..	..
Viet Nam	..	..	12	..	..	..	0.13	..
Europe and Central Asia								
Albania	..	..	..	..	..	..	..	..
Armenia			..	..			..	..
Azerbaijan			..	..			..	..
Belarus			..	..			..	..
Bosnia and Herzegovina			..	..			..	..
Bulgaria	..	..	..	..	..	..	..	..
Georgia			..	..			..	..
Kazakhstan			..	..			..	..
Kyrgyzstan			..	..			..	..
Latvia			..	..			..	..
Lithuania			..	..			..	..

TABLE A6 *(cont.)*

	Public expenditures on agricultural research and development							
	Total *(Million constant 2005 PPP dollars)*				As a share of agricultural GDP *(Percentage)*			
	1981	1990	2000	Latest year	1981	1990	2000	Latest year
Montenegro				..				..
Republic of Moldova			..	..			..	..
Romania	..	..	..	..	..	..	..	..
Russian Federation			..	..			..	..
Serbia				..				..
Serbia and Montenegro			..				..	
Tajikistan			..	..			..	..
The former Yugoslav Republic of Macedonia			..	..			..	..
Turkey	..	..	..	..	..	..	..	..
Turkmenistan			..	..			..	..
Ukraine			..	..			..	..
USSR	..	..			..	..		
Uzbekistan			..	..			..	..
Yugoslav SFR	..	..			..	..		
Latin America and the Caribbean								
Antigua and Barbuda	..	..	..	..	..	..	..	..
Argentina	203	194	239	449	1.17	0.98	1.34	1.27
Belize	1	1	2	3	0.75	0.86	0.90	0.95
Bolivia (Plurinational State of)	..	..	..	..	..	..	..	..
Brazil	979	1 227	1 247	1 296	1.15	1.66	1.86	1.68
Chile	58	75	117	98	1.45	1.09	1.30	1.22
Colombia	104	153	165	152	0.43	0.54	0.62	0.50
Costa Rica	13	17	25	30	0.41	0.85	0.93	0.93
Cuba	..	..	..	..	..	..	..	..
Dominica	..	..	..	..	..	..	..	..
Dominican Republic	..	..	..	17	..	..	..	0.26
Ecuador	..	..	..	..	..	..	..	..
El Salvador	14	11	7	6	0.20	0.30	0.20	0.15
Grenada	..	..	..	..	..	..	..	..
Guatemala	21	14	9	8	0.25	0.15	0.07	0.06
Guyana	..	..	..	..	..	..	..	..
Haiti	..	..	..	..	..	..	..	..
Honduras	6	15	14	13	0.25	0.55	0.54	0.43
Jamaica	..	..	..	..	..	..	..	..
Mexico	..	..	438	518	..	..	1.08	1.21
Nicaragua	..	..	..	24	..	..	..	0.94
Panama	10	12	11	10	0.92	0.95	0.72	0.50
Paraguay	..	..	..	3	..	..	..	0.20
Peru	..	..	..	..	..	..	..	..

TABLE A6 *(cont.)*

	Public expenditures on agricultural research and development							
	Total (*Million constant 2005 PPP dollars*)				As a share of agricultural GDP (*Percentage*)			
	1981	1990	2000	Latest year	1981	1990	2000	Latest year
Saint Lucia	..	..	..	..	..	..	..	..
Saint Vincent and the Grenadines	..	..	..	..	..	..	..	..
Suriname	..	..	..	..	..	..	..	..
Uruguay	18	29	38	60	0.67	1.45	2.06	1.99
Venezuela (Bolivarian Republic of)	..	..	..	..	..	..	..	..
Middle East and North Africa								
Algeria	..	..	..	..	..	..	..	..
Djibouti	..	..	..	..	..	..	..	..
Egypt	..	..	..	..	..	..	..	..
Iran (Islamic Republic of)	..	..	508	559	..	..	0.76	0.82
Iraq	..	..	..	..	..	..	..	..
Jordan	..	..	7	..	..	..	1.99	..
Lebanon	..	..	..	..	..	..	..	..
Libya	..	..	..	..	..	..	..	..
Morocco	99	119	105	..	1.72	1.01	1.00	..
Occupied Palestinian Territory	..	..	..	..	..	..	..	..
Syrian Arab Republic	..	..	79	..	..	..	0.53	..
Tunisia	..	..	45	..	..	..	0.71	..
Yemen	..	..	..	..	..	..	..	..
South Asia								
Afghanistan	..	..	..	..	..	..	..	..
Bangladesh	..	..	142	126	..	..	0.46	0.31
Bhutan	..	..	..	..	..	..	..	..
India	414	714	1 487	2 276	0.22	0.29	0.39	0.40
Maldives	..	..	..	..	..	..	..	..
Nepal	..	..	25	22	..	..	0.29	0.23
Pakistan	..	..	136	172	..	..	0.21	0.21
Sri Lanka	..	..	55	38	..	..	0.54	0.34
Sub-Saharan Africa								
Angola	..	..	..	..	..	..	..	..
Benin	6	11	13	22	0.44	0.57	0.43	0.57
Botswana	9	11	20	19	1.94	2.50	4.50	4.32
Burkina Faso	23	22	23	19	1.66	1.23	0.79	0.43
Burundi	..	..	4	10	..	..	0.45	1.78
Cameroon	..	..	..	..	..	..	..	..
Cape Verde	..	..	..	..	..	..	..	..
Central African Republic	..	..	..	..	..	..	..	..
Chad	..	..	..	..	..	..	..	..

TABLE A6 *(cont.)*

	Public expenditures on agricultural research and development							
	Total (Million constant 2005 PPP dollars)				As a share of agricultural GDP (Percentage)			
	1981	1990	2000	Latest year	1981	1990	2000	Latest year
Comoros	..	..	..	..	..	..	..	..
Congo	..	..	3	5	..	..	0.60	0.85
Côte d'Ivoire	72	74	56	43	1.17	0.95	0.77	0.54
Democratic Republic of the Congo	..	..	..	..	..	..	..	..
Eritrea			9	3			2.53	0.45
Ethiopia			49	69			0.31	0.27
Ethiopia PDR	17	39			0.14	0.28		
Gabon	..	..	2	2	..	..	0.24	0.20
Gambia	..	..	3	3	..	..	0.58	0.50
Ghana	14	40	41	95	0.25	0.66	0.57	0.90
Guinea	..	..	10	4	..	..	0.73	0.18
Guinea-Bissau	..	..	..	..	..	..	..	..
Kenya	88	127	151	172	1.36	1.50	1.31	1.30
Lesotho	..	..	..	..	..	..	..	..
Liberia	..	..	..	..	..	..	..	..
Madagascar	14	21	9	12	0.48	0.70	0.24	0.27
Malawi	29	39	30	21	1.84	1.71	1.03	0.68
Mali	33	24	31	25	1.56	0.95	0.95	0.57
Mauritania	..	..	..	6	..	..	..	1.16
Mauritius	11	12	23	22	2.18	1.69	3.41	3.92
Mozambique	..	..	..	18	..	..	..	0.38
Namibia	..	..	..	22	..	..	..	2.03
Niger	9	15	5	6	0.37	0.81	0.19	0.17
Nigeria	231	117	191	404	0.40	0.13	0.21	0.42
Rwanda	..	..	..	18	..	..	..	0.53
Sao Tome and Principe	..	..	..	..	..	..	..	..
Senegal	41	34	25	25	2.36	1.78	1.02	0.87
Seychelles	..	..	..	..	..	..	..	..
Sierra Leone	..	..	..	6	..	..	..	0.31
Somalia	..	..	..	..	..	..	..	..
South Africa	221	247	283	272	1.44	2.09	2.83	2.02
Sudan	54	29	37	52	0.73	0.28	0.20	0.27
Swaziland	..	..	..	..	..	..	..	..
Togo	16	13	13	9	1.97	1.11	0.88	0.47
Uganda	..	..	40	88	..	..	0.76	1.24
United Republic of Tanzania	..	..	44	77	..	..	0.36	0.50
Zambia	27	23	15	8	1.81	1.29	0.71	0.29
Zimbabwe	..	..	..	..	..	..	..	..

TABLE A7

Official development assistance to agriculture and agricultural share of ODA to all sectors, 1980, 1990, 2000 and 2009

	Official development assistance							
	To agriculture (Million constant 2005 US$)				Agricultural share of ODA to all sectors (Percentage)			
	1980	1990	2000	2010	1980	1990	2000	2010
WORLD	**8 397**	**8 193**	**4 131**	**8 299**	**18.8**	**14.5**	**5.6**	**5.9**
LOW- AND MIDDLE-INCOME COUNTRIES	**8 328**	**8 150**	**4 119**	**8 266**	**20.0**	**15.2**	**5.6**	**5.8**
East Asia and the Pacific	**1 358**	**1 851**	**722**	**728**	**18.0**	**17.3**	**5.3**	**5.9**
American Samoa	..	..	..	..	..	..	..	..
Cambodia	6	..	59	50	7.6	..	10.8	5.6
China	..	1 096	193	65	..	53.7	5.7	3.1
Democratic People's Republic of Korea	..	36	1	0	..	..	1.4	0.8
Fiji	5	1	1	12	35.4	6.9	6.7	20.5
Indonesia	704	400	132	170	23.6	11.7	6.1	5.9
Kiribati	5	1	10	2	37.9	7.9	44.1	5.3
Lao People's Democratic Republic	70	48	18	18	56.2	20.6	6.2	3.4
Malaysia	..	15	6	6	..	2.1	0.4	6.7
Marshall Islands	..	..	5	1	..	..	8.3	0.8
Micronesia (Federated States of)	..	..	14	1	..	..	12.1	1.0
Mongolia	..	..	5	26	..	..	1.6	4.9
Myanmar	136	..	3	22	28.0	..	3.5	7.5
Palau	..	..	0	0	..	..	0.9	2.3
Papua New Guinea	11	2	25	40	1.7	1.1	4.5	4.8
Philippines	133	145	157	49	18.7	6.0	13.0	4.2
Samoa	14	4	2	2	73.7	7.4	4.1	1.6
Solomon Islands	6	5	3	17	14.5	17.5	2.9	5.1
Thailand	227	73	22	11	31.8	5.1	1.7	2.2
Timor-Leste	..	..	5	10	..	..	1.7	2.8
Tonga	1	4	0	1	7.2	9.8	0.5	1.5
Tuvalu	1	..	..	0	11.3	..	0.0	1.9
Vanuatu	5	2	3	2	5.4	15.8	10.1	2.6
Viet Nam	35	20	58	223	9.4	10.1	2.9	6.4
Europe and Central Asia			**272**	**488**			**5.2**	**6.1**
Albania	..	..	10	16	..	..	2.3	3.8
Armenia			15	6			5.6	1.8
Azerbaijan			60	3			28.1	1.6
Belarus			..	0			..	0.2
Bosnia and Herzegovina			15	10			1.7	1.9
Bulgaria	..	..	..	..	..	..	..	..
Georgia			21	22			7.3	2.8
Kazakhstan			3	2			1.0	0.6
Kyrgyzstan			75	9			22.4	1.9

TABLE A7 *(cont.)*

	Official development assistance							
	To agriculture *(Million constant 2005 US$)*				Agricultural share of ODA to all sectors *(Percentage)*			
	1980	1990	2000	2010	1980	1990	2000	2010
Latvia			..	..			..	..
Lithuania			..	..			..	..
Montenegro				4				3.9
Republic of Moldova			11	97			8.1	12.8
Romania	..	..	..	..	..	..	..	..
Russian Federation			..	..			..	..
Serbia				25				4.2
Serbia and Montenegro			15				0.7	
Tajikistan			22	26			15.5	6.3
The former Yugoslav Republic of Macedonia			24	21			7.9	9.7
Turkey	..	1	1	238	..	0.3	0.1	15.5
Turkmenistan			0	1			0.2	2.5
Ukraine			..	3			..	0.5
USSR	..	..			..	..		
Uzbekistan			0	5			0.2	0.7
Yugoslav SFR	..	..			..	..		
Latin America and the Caribbean	**772**	**665**	**522**	**960**	**20.3**	**10.1**	**6.5**	**8.5**
Antigua and Barbuda	..	..	10	0	..	..	90.1	2.1
Argentina	..	25	2	34	..	14.2	2.5	25.0
Belize	1	5	6	22	14.2	37.6	27.1	46.8
Bolivia (Plurinational State of)	19	174	115	122	12.9	23.5	9.1	18.1
Brazil	30	20	19	201	8.4	19.2	8.3	22.4
Chile	2	0	3	2	10.9	0.1	5.8	1.7
Colombia	..	3	82	100	..	2.6	5.1	9.1
Costa Rica	41	15	13	2	28.8	9.9	22.0	2.5
Cuba	..	0	7	8	..	1.0	11.0	7.5
Dominica	4	0	8	0	30.8	0.0	47.0	0.3
Dominican Republic	25	6	12	17	14.8	4.2	4.0	4.3
Ecuador	0	10	10	38	0.0	2.3	4.4	15.7
El Salvador	165	32	17	10	57.7	7.6	11.2	3.5
Grenada	0	..	2	0	6.7	..	14.5	0.7
Guatemala	..	24	24	41	..	8.5	8.6	9.9
Guyana	6	44	0	22	28.6	14.2	0.1	7.6
Haiti	6	35	23	68	6.8	13.3	11.9	2.1
Honduras	45	82	59	66	10.4	17.0	7.4	12.2
Jamaica	25	68	7	25	15.3	26.1	5.9	24.8
Mexico	1	1	5	20	1.3	0.2	1.3	2.0
Nicaragua	142	67	45	59	27.9	8.2	8.1	10.6
Panama	11	..	1	4	14.1	..	2.1	8.3

TABLE A7 *(cont.)*

	Official development assistance							
	To agriculture *(Million constant 2005 US$)*				Agricultural share of ODA to all sectors *(Percentage)*			
	1980	1990	2000	2010	1980	1990	2000	2010
Paraguay	17	14	3	20	9.1	9.4	5.9	10.5
Peru	227	31	24	72	51.1	16.1	2.3	10.8
Saint Lucia	2	..	10	0	25.0	..	37.5	4.0
Saint Vincent and the Grenadines	..	0	8	0	..	100.0	57.7	3.0
Suriname	2	6	4	..	2.2	18.1	15.6	0.0
Uruguay	..	2	1	2	..	8.2	8.1	6.4
Venezuela (Bolivarian Republic of)	..	..	0	0	..	..	0.2	0.8
Middle East and North Africa	**742**	**345**	**316**	**492**	**13.1**	**6.9**	**5.4**	**4.0**
Algeria	..	28	1	8	..	11.8	0.9	3.2
Djibouti	4	11	1	1	19.7	25.8	1.1	0.8
Egypt	325	108	155	133	9.3	4.1	8.5	6.0
Iran (Islamic Republic of)	..	..	0	2	..	..	0.1	1.7
Iraq	..	..	..	51	..	..	0.0	2.5
Jordan	24	4	37	2	8.2	0.5	5.5	0.2
Lebanon	..	..	7	27	..	..	6.5	6.4
Libya	..	1	..	0	..	33.8	..	0.1
Morocco	19	72	6	161	7.8	10.7	0.8	8.7
Occupied Palestinian Territory	..	..	17	34	..	..	2.8	1.6
Syrian Arab Republic	110	..	8	26	94.9	..	7.0	6.1
Tunisia	163	83	51	4	18.2	21.4	8.1	0.6
Yemen	98	38	31	42	20.7	15.4	5.2	5.2
South Asia	**3 336**	**1 924**	**504**	**1 085**	**28.6**	**21.0**	**7.9**	**4.8**
Afghanistan	..	33	2	583	..	28.1	1.2	8.7
Bangladesh	797	590	177	121	23.4	25.1	11.2	5.0
Bhutan	1	18	4	6	3.3	46.9	5.0	5.7
India	2 000	237	194	189	37.3	8.0	7.8	2.8
Maldives	..	0	0	0	..	0.1	0.0	0.1
Nepal	141	196	61	30	36.9	52.6	16.6	2.9
Pakistan	201	755	12	145	15.6	37.2	0.9	3.3
Sri Lanka	197	94	55	12	16.4	7.6	14.0	1.2
Sub-Saharan Africa	**2 082**	**2 897**	**1 488**	**2 857**	**19.6**	**16.0**	**7.1**	**7.4**
Angola	16	51	9	27	14.5	17.9	2.5	6.1
Benin	1	16	45	21	0.6	8.1	10.9	3.6
Botswana	15	3	2	1	9.6	2.6	4.4	0.6
Burkina Faso	98	66	127	71	31.8	21.0	23.8	8.7
Burundi	33	68	11	96	12.7	28.2	6.3	19.2
Cameroon	133	142	16	62	62.7	21.6	3.1	9.4
Cape Verde	17	5	4	3	19.8	6.0	4.8	1.7

TABLE A7 *(cont.)*

	Official development assistance							
	To agriculture (Million constant 2005 US$)				Agricultural share of ODA to all sectors (Percentage)			
	1980	1990	2000	2010	1980	1990	2000	2010
Central African Republic	5	41	2	3	4.4	16.6	1.4	1.2
Chad	1	21	27	9	4.5	11.2	6.2	1.8
Comoros	11	5	2	2	31.3	49.1	6.0	2.2
Congo	7	33	1	1	4.3	13.8	1.4	0.1
Côte d'Ivoire	66	98	37	93	52.1	11.1	6.4	12.8
Democratic Republic of the Congo	103	43	3	100	21.7	5.9	1.3	3.1
Eritrea			39	13			9.2	13.6
Ethiopia			67	222			5.3	7.3
Ethiopia PDR	62	93			30.0	12.4		
Gabon	32	1	10	28	62.5	5.6	11.6	15.4
Gambia	5	12	4	15	6.1	12.1	7.9	9.1
Ghana	143	60	140	126	38.7	6.7	16.4	8.1
Guinea	28	77	9	8	16.6	16.6	4.4	6.0
Guinea-Bissau	4	37	0	10	8.2	24.7	0.2	8.3
Kenya	175	246	78	323	24.1	13.0	6.1	10.6
Lesotho	24	9	4	0	12.9	12.5	6.9	0.1
Liberia	10	2	4	36	19.0	12.2	13.8	2.6
Madagascar	9	45	50	19	3.2	7.9	9.6	6.5
Malawi	35	113	50	79	19.6	23.0	6.5	8.6
Mali	40	116	137	223	24.7	34.1	19.4	24.9
Mauritania	24	75	27	11	20.9	34.6	12.0	3.5
Mauritius	1	4	2	0	0.9	4.5	6.5	0.0
Mozambique	95	209	48	72	39.0	21.7	2.6	3.7
Namibia	..	4	13	14	..	4.2	9.9	4.0
Niger	70	115	81	43	25.2	51.1	20.7	8.1
Nigeria	11	11	16	69	75.5	2.2	1.6	5.5
Rwanda	54	26	62	68	20.9	10.1	10.1	7.5
Sao Tome and Principe	2	3	6	3	39.7	6.4	12.3	9.0
Senegal	51	104	60	256	16.6	11.5	7.2	19.7
Seychelles	4	7	0	1	32.0	20.4	5.3	2.0
Sierra Leone	1	..	1	75	1.6	..	0.3	16.7
Somalia	122	56	0	1	29.9	21.6	0.2	0.3
South Africa	..	..	12	20	..	..	2.5	1.9
Sudan	351	55	1	145	22.2	20.1	0.2	7.5
Swaziland	0	5	5	6	0.7	26.0	15.0	5.0
Togo	41	33	6	36	19.4	13.1	6.5	7.6
Uganda	5	211	88	198	1.5	23.5	6.6	10.8
United Republic of Tanzania	124	497	127	132	10.1	30.0	7.5	4.7
Zambia	51	47	39	36	10.8	5.5	2.9	4.9
Zimbabwe	2	32	17	81	0.6	8.9	7.2	15.5

TABLE A7 *(cont.)*

	Official development assistance							
	To agriculture *(Million constant 2005 US$)*				Agricultural share of ODA to all sectors *(Percentage)*			
	1980	1990	2000	2010	1980	1990	2000	2010
HIGH-INCOME COUNTRIES	**64**	**75**	**10**	**35**	**2.6**	**2.6**	**6.3**	**9.2**
Andorra	..	..	..	..	..	..	..	..
Aruba	..	0	..	..	..	0.3	..	..
Australia	..	..	..	..	..	..	..	..
Austria	..	..	..	..	..	..	..	..
Bahamas	..	..	..	..	..	..	..	..
Bahrain	..	..	0	..	..	..	0.9	..
Barbados	9	0	0	0	..	10.5	0.1	1.7
Belgium			..	..			..	..
Belgium-Luxembourg	..	..			..	..		
Bermuda	..	..	..	..	..	..	..	..
Brunei Darussalam	..	..	..	..	..	..	..	..
Canada	..	..	..	..	..	..	..	..
Cayman Islands	..	..	..	..	..	..	..	..
China, Hong Kong SAR	..	..	..	..	..	..	..	..
China, Macao SAR	..	..	..	..	..	..	..	..
Croatia	..	..	1	34	..	..	0.7	16.7
Cyprus	..	..	..	..	..	..	..	..
Czech Republic			..	..			..	..
Czechoslovakia	..	..			..	..		
Denmark	..	..	..	..	..	..	..	..
Equatorial Guinea	..	19	1	0	..	59.4	5.5	0.5
Estonia			..	..			..	..
Faroe Islands	..	..	..	..	..	..	..	..
Finland	..	..	..	..	..	..	..	..
France	..	..	..	..	..	..	..	..
Germany	..	..	..	..	..	..	..	..
Greece	..	..	..	..	..	..	..	..
Greenland	..	..	..	..	..	..	..	..
Guam	..	..	..	..	..	..	..	..
Hungary	..	..	..	..	..	..	..	..
Iceland	..	..	..	..	..	..	..	..
Ireland	..	..	..	..	..	..	..	..
Israel	45	1	..	..	2.4	0.0	..	..
Italy	..	..	..	..	..	..	..	..
Japan	..	..	..	..	..	..	..	..
Kuwait	..	..	..	..	..	..	..	..
Liechtenstein	..	..	..	..	..	..	..	..
Luxembourg			..	..			..	..
Malta	2	..	..	..	79.8	..	0.0	..
Monaco	..	..	..	..	..	..	..	..

TABLE A7 *(cont.)*

	Official development assistance							
	To agriculture *(Million constant 2005 US$)*				Agricultural share of ODA to all sectors *(Percentage)*			
	1980	**1990**	**2000**	**2010**	**1980**	**1990**	**2000**	**2010**
Netherlands	..	..	..	..	..	..	..	..
New Caledonia	4	2	..	..	18.0	3.7	..	..
New Zealand	..	..	..	..	..	..	..	..
Northern Mariana Islands	..	6	..	..	..	36.0	..	..
Norway	..	..	..	..	..	..	..	..
Oman	..	11	8	0	..	68.1	67.2	0.3
Poland	..	..	..	..	..	..	..	..
Portugal	..	..	..	..	..	..	..	..
Puerto Rico	..	..	..	..	..	..	..	..
Qatar	..	..	..	..	..	..	..	..
Republic of Korea	..	36	..	..	..	..	..	..
San Marino	..	..	..	..	..	..	0.6	..
Saudi Arabia	..	..	0	..	..	..	..	..
Singapore	4	..	..	..	27.6	..	..	..
Slovakia			..	..			0.0	..
Slovenia			0	..			..	..
Spain	..	..	..	..	..	..	..	..
Sweden	..	..	..	..	..	..	..	..
Switzerland	..	..	..	..	..	..	0.4	0.4
Trinidad and Tobago	..	..	0	0	..	..	1.7	..
Turks and Caicos Islands	..	..	0	..	..	..	..	..
United Arab Emirates	..	..	..	..	..	..	..	..
United Kingdom	..	..	..	..	..	..	..	..
United States of America	..	..	..	..	..	..	..	..
United States Virgin Islands	..	..	..	..	..	..	..	..
Unspecified recipients	**8**	**206**	**125**	**1 113**	**1.7**	**13.2**	**1.6**	**4.5**
Regional recipients	**32**	**262**	**169**	**543**	**11.0**	**16.0**	**5.0**	**6.0**

- **References**
- **Special chapters of**
  The State of Food and Agriculture

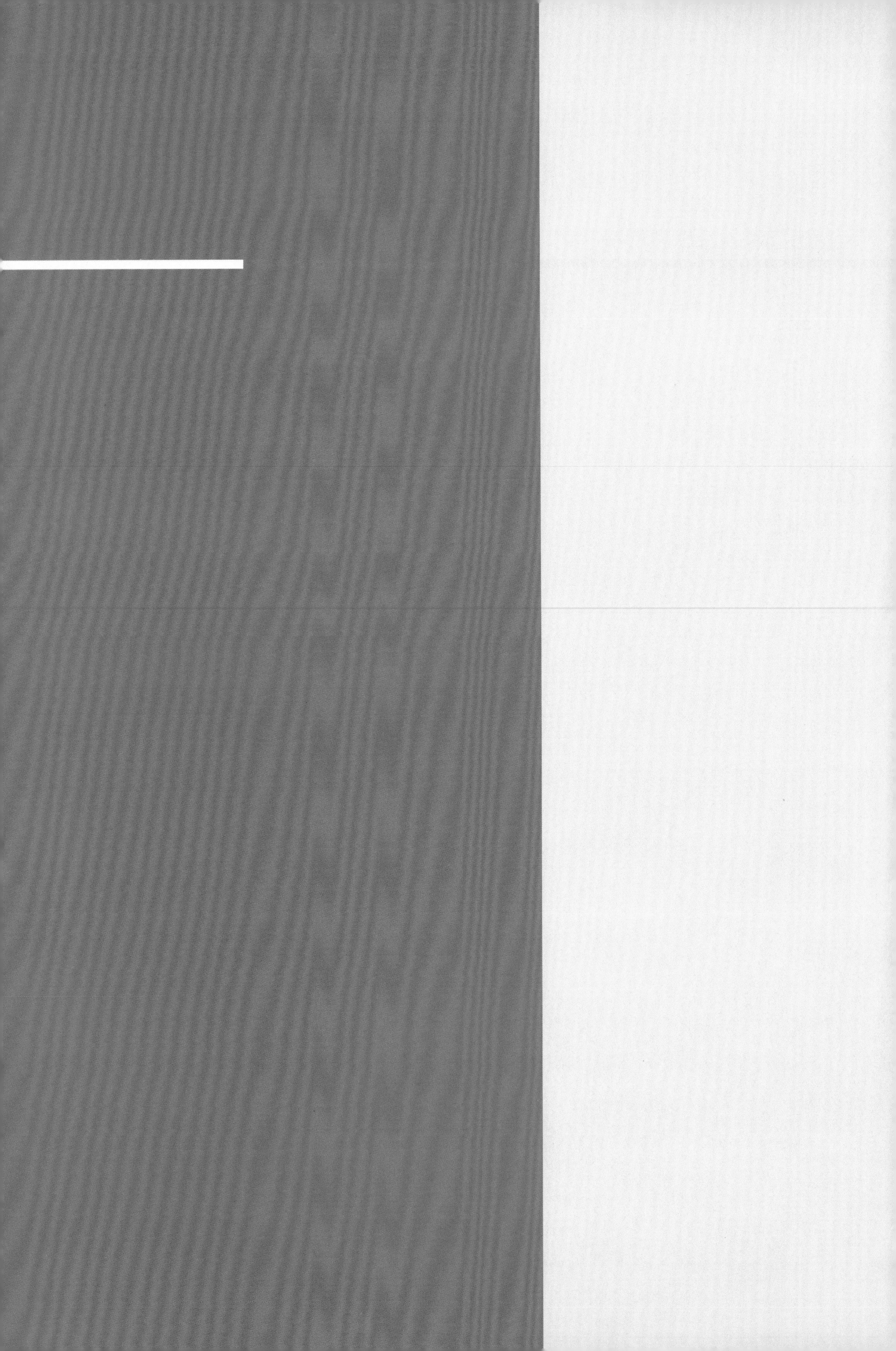

References

Ahmed, A.U., Rabbani, M., Sulaiman, M. & Das, N.C. 2009. *The impact of asset transfer on livelihoods of the ultra poor in Bangladesh.* Research Monograph Series No. 39. Dhaka, Bangladesh Rural Advancement Committee.

Ahmed, R. & Hossain, M. 1990. *Development impact of rural infrastructure in Bangladesh.* IFPRI Research Report 83. Washington, DC, IFPRI.

Akroyd, S. & Smith, L. 2007. *Review of public spending to agriculture.* A joint DFID/ World Bank study. Oxford, UK, Oxford Policy Management.

Allcott, H., Lederman, D. & López, R. 2006. *Political institutions, inequality, and agricultural growth: the public expenditure connection.* Policy Research Working Paper Series 3902. Washington, DC, World Bank.

Alston, J. 2010. *The benefits from agricultural research and development, innovation, and productivity growth.* OECD Food, Agriculture and Fisheries Working Papers No. 31. Paris, OECD.

Alston, J.M., Beddow, J.M. & Pardey, P.G. 2010. Global patterns of crop yields and other partial productivity measures and prices. *In* J.M. Alston, B.A. Babcock & P.G. Pardey, eds. *The shifting patterns of agricultural productivity worldwide.* CARD-MATRIC Electronic Book. Ames, USA, Center for Agricultural and Rural Development, The Midwest Agribusiness Trade Research and Information Center, Iowa State University (available at www.matric.iastate.edu/shifting_patterns).

Alston, J., Marra, M., Pardey, P. & Wyatt, T. 2000. Research returns redux: a meta-analysis of the returns to agricultural R&D. *Australian Journal of Agricultural and Resource Economics,* 44(2): 185–215.

Anderson, K., ed. 2009. *Distortions to agricultural incentives: a global perspective, 1955–2007.* London, Palgrave Macmillan and Washington, DC, World Bank.

Anderson, K. & Brückner, M. 2011. *Price distortions and economic growth in Sub-Saharan Africa.* CEPR (Center for Economic Policy Research) Discussion Papers 8530. London, CEPR.

Anderson, K. & Nelgen, S. 2012. *Updated national and global estimates of distortions to agricultural incentives, 1955 to 2010,* Washington, DC, World Bank.

Anderson, K. & Valenzuela, E. 2008. *Estimates of global distortions to agricultural incentives, 1955 to 2007.* Washington, DC, World Bank.

Anderson, K., Lloyd, P. & MacLaren, D. 2007. Distortions to agricultural incentives in Australia since World War II. *The Economic Record,* 83(263): 461–482.

Anderson, K., Valenzuela, E. & van der Mensbrugghe, D. 2009. *Welfare and poverty effects of global agricultural and trade policies using the linkage model.* Agricultural Distortions Working Paper 52785. Washington, DC, World Bank.

Anseeuw, W., Ducastel, A. & Gabas, J. 2011. *The end of the African peasant? From investment funds and finance value chains to peasant related questions.* Paper presented at the International Conference on Global Land Grabbing. Brighton, UK, 6–8 April 2011.

Anseeuw, W., Wily, L.A., Cotula, L. & Taylor, M. 2012. *Land rights and the rush for land: findings of the global commercial pressures on land research project.* Rome, ILC (International Land Coalition).

Anson, R. & Zegarra, E. 2008. *Honduras: public expenditure assessment and strategy for an enhanced agricultural and forestry sector.* Draft paper prepared for the World Bank Agriculture and Rural Development Sector in collaboration with RUTA. Washington, DC, World Bank.

Antle, J.M. 1983. Infrastructure and aggregate agricultural productivity: international evidence. *Economic Development and Cultural Change,* 31(3): 609–619.

Arezki, R., Deininger, K. & Selod, H. 2011. *What drives the global land rush?* IMF Working Papers 11/251. Washington, DC, International Monetary Fund.

Arslan, A., McCarthy, N., Lipper, L., Asfaw, S. & Cattaneo, A. 2012 (forthcoming). *Adoption and intensity of adoption of conservation agriculture in Zambia.* ESA Working paper. Rome, FAO.

Augusto, H.A. & Ribeiro, E.M. 2006. *O Idoso Rural e os Efeitos das Aprosentadorias Rurais nos Domicílios e no Comercio Local: O Caso de Medina, Nordeste de Minas.* Paper presented at the meetings of the Associação Brasileira de

Estudos Populacionais, Caxambu, Brazil, 18–22 September. (mimeo)

Ballard, R. 1987. The political economy of migration: Pakistan, Britain, and the Middle East. *In* J. Eades, ed. *Migrants, workers and the social order.* London, Tavistock.

Banerjee, A.V. & Duflo, E. 2004. *Growth theory through the lens of development economics.* Working Paper 05-01. Cambridge, USA, Massachusetts Institute of Technology.

Barrett, C.B., Bellemare, M.F. & Hou, J.Y. 2010. Reconsidering conventional explanations of the inverse productivity-size relationship. *World Development,* 38(1): 88–97.

Barrientos, A. 2011. Social transfers and growth: what do we know? What do we need to find out? *World Development*, 40(1):11–20.

Barrientos, A., Ferreira, M., Gorman, M., Heslop, A., Legido-Quigley, H., Lloyd-Sherlock, P., Møller, V., Saboia, J. & Werneck, M.L.T. 2003. *Non-contributory pensions and poverty prevention: a comparative study of South Africa and Brazil.* London, HelpAge International and Institute for Development Policy and Management.

Baxter, J. 2011. *Understanding land investment deals in Africa. Country report: Mali.* Oakland, CA, USA, The Oakland Institute.

BBC (British Broadcasting Corporation). 2012. *Cambodia suspends new land grants for companies.* Online news story, 7 May (available at http://www.bbc.co.uk/news/world-asia-17980399).

Becker, G. 1983. A theory of competition among pressure groups for political influence. *Quarterly Journal of Economics,* 98(3): 371–400.

Beintema, N.M. & Elliott, H. 2011. Setting meaningful investment targets in agricultural research and development: challenges, opportunities and fiscal realities. *In* P. Conforti, ed. *Looking ahead in world food and agriculture: perspectives to 2050.* Rome, FAO.

Beintema, N.M. & Stads G.J. 2008a. *Diversity in agricultural research resources in the Asia–Pacific region.* Bangkok, Asia–Pacific Association of Agricultural Research Institutions (APAARI) and Washington, DC, IFPRI.

Beintema, N.M. & Stads, G.J., 2008b. *Measuring agricultural research investments: a revised global picture.* Agriculture Science and Technology Indicators (ASTI) Background Note. Washington, DC, IFPRI.

Benin, S., Nin-Pratt, A. & Randriamamonjy, J. 2007. *Agricultural productivity growth and government spending in sub-Saharan Africa.* Washington, DC, IFPRI. (mimeo)

Benin, S., Nkonya, E., Okecho, G., Randriamamonjy, J., Kato, E., Lubade, G. & Kyotalimye, M. 2011. Returns to spending on agricultural extension: the case of the National Agricultural Advisory Services (NAADS) programme of Uganda. *Agricultural Economics,* 42(2): 249–267.

Berhane, G., Hoddinott, J., Kumar, N. & Taffesse, A.S. 2011. *The impact of Ethiopia's productive safety nets and household asset building programme: 2006–2010.* Washington, DC, IFPRI.

Bezemer, D. & Headey, D. 2008. Agriculture, development, and urban bias. *World Development*, 36(8): 1342–1364.

Binswanger, H.P. 1983. *Growth and employment in rural Thailand.* World Bank Report No. 3906-TH. Washington, DC, World Bank.

Binswanger, H.P., Khandker, S.R. & Rosenzweig, M.R. 1993. How infrastructure and financial institutions affect agricultural output and investment in India. *Journal of Development Economics,* 41(2): 337–366. Amsterdam, Elsevier.

Bioversity, CGIAR Consortium, FAO, IFAD, IFPRI, IICA, OECD, UNCTAD, Coordination Team of UN High Level Task Force on the Food Security Crisis, WFP, World Bank & WTO. 2012. *Sustainable agricultural productivity growth and bridging the gap for small family farms.* Interagency report to the Mexican G20 Presidency (available at http://www.fao.org/economic/g20/en/).

Birner, R. & Resnick, D. 2010. The political economy of policies for smallholder agriculture. *World Development,* 38(10): 1442–1452.

Böber, C. 2012. *The determinants of farm investment of Nepalese households: a case-study on the relationship between credit access and the variation in productive agricultural capital at the farm level.* Background paper for *The State of Food and Agriculture 2012.* Rome, FAO.

Boone, R., Covarrubias, K., Davis, B. & Winters, P. 2012. *Cash transfer programs and agricultural production: the case of Malawi.* Rome, FAO. (mimeo)

Bouis, H., Graham, R. & Welch, R. 2000. The CGIAR micronutrients project: justification and objectives. *Food and Nutrition Bulletin,* 21(4): 374–381.

Buckwell, A. 2005. **Green accounting for agriculture**. *Journal of Agricultural Economics,* 56(2):187–215.

Byerlee, D., de Janvry, A. & Sadoulet, E. 2009. Agriculture for development: toward a new paradigm. *Annual Review of Resource Economics*, 1(1): 15–31.

Cammack, T., Fowler, M. & Phomdouangsy, C.D. 2008. *Lao PDR public expenditure study*. Public Expenditures for Pro-Poor Agricultural Growth. Department for International Development (DFID) /World Bank (ARD) Partnership. Draft.

CDRI (Cambodia Development Resource Institute). 2011. *Foreign investment in agriculture in Cambodia*. Presented at FAO's Expert Meeting on International Investment in the Agriculture Sector of Developing Countries. Rome, Italy, 22–23 November, 2011.

Christiaensen, L. & Demery, L. 2007. *Down to earth: agriculture and poverty reduction in Africa*. Washington, DC, World Bank.

Christiaensen, L., Demery, L. & Kuhl, J. 2010. The (evolving) role of agriculture in poverty reduction: an empirical perspective. *Journal of Development Economics*, 96: 239–254.

Christy, R., Mabaya, E., Wilson, N., Mutambatsere, E. & Mhlang, N. 2009. Enabling, environments for competitive agro-industries. *In* C.A. da Silva, D. Baker, A.W. Shepard, C. Jenane and S. Miranda-da-Cruz, eds. *Agro-industries for development*, pp.136–85. Rome, FAO and UNIDO (United Nations Industrial Development Organization).

Claessens, S. 2005. *Access to financial services: a review of the issues and public policy objectives.* Policy Research Working Paper Series 3589. Washington, DC, World Bank.

Coate, S. & Morris, S. 1999. Policy persistence. *American Economic Review*, 89(5): 1327–1336.

Cotula, L. & Polack, E. 2012. *The global land rush: what the evidence reveals about scale and geography*. IIED (International Institute for Environment and Development) Briefing. London, IIED.

Cotula, L., Vermeulen, S., Leonard, R. & Keeley, J. 2009. *Land grab or development opportunity? Agricultural investment and international land deals in Africa*. Rome and London, FAO, IFAD and IIED.

Covarrubias, K., Davis, B. & Winters, P. 2012. From protection to production: productive impacts of the Malawi social cash transfer scheme. *Journal of Development Effectiveness*, 4(1): 50–77.

Crego, A., Larson, D., Butzer, R. & Mundlak, Y. 1997. *A new database on investment and capital for agriculture and manufacturing.* Policy Research Working Paper No. 2013. Washington, DC, World Bank.

Cuffaro, N. & Hallam, D. 2011. *"Land grabbing" in developing countries: foreign investors, regulation and codes of conduct*. Paper presented at the International Conference on Global Land Grabbing. Brighton, UK, 6–8 April 2011.

da Silva, C.A., Baker, D., Shepard, A.W., Jenane, C. & Miranda-da-Cruz, S. 2009. *Agro-industries for development.* Rome, FAO and UNIDO.

Daidone, S. & Anríquez, G. 2011. *An extended cross-country database for agricultural investment and capital.* ESA Working Paper No. 11–16. Rome, FAO.

Daley, E. & Park, C.M. 2011. *The gender and equity implications of land-related investments on labour and income-generating opportunities. A case study of agricultural investments in Northern Tanzania.* Final Report. Rome, FAO.

Dastagiri, M.B. 2010. The effect of government expenditure on promoting livestock GDP and reducing rural poverty in India. *Outlook on Agriculture*, 39(2): 127–133.

Datt, G. & Ravallion, M. 1998. Farm productivity and rural poverty in India. *Journal of Development Studies*, 34(4): 62–85.

Davies, G. 2011. *Farmland as an asset class: the focus of private equity firms in Africa.* Paper presented at the International Conference on Global Land Grabbing. Brighton, UK, 6–8 April 2011.

Dayal, H. & Karan, A.K. 2003. *Labour migration from Jharkhand*. New Delhi, Institute for Human Development.

de Brauw, A. & Rozelle, S. 2008. Migration and household investment in rural China. *China Economic Review*, 19: 320–335.

de Haas, H. 2007. *Migration and development: A theoretical perspective*. International Migration Institute Working Paper No. 9. Oxford, UK, International Migration Institute, University of Oxford.

de Janvry, A. 2009. Annex: agriculture for development – implications for agro-industries. *In* C.A. da Silva, D. Baker, A.W. Shepard, C. Jenane and S. Miranda-da-Cruz, eds. *Agro-industries for development*, pp.252–270. Rome, FAO and UNIDO.

de la Croix, D. & Delavallade, C. 2009. Growth, public investment, and corruption with failing institutions. *Economics of Governance*, 10(3): 187–219.

Deacon, R.T. 2003. *Dictatorship, democracy, and the provision of public goods*. Economics

Working Paper Series 11925. Santa Barbara, CA, USA, University of Santa Barbara.

de Gorter, H. & Just, D. 2010. The social costs and benefits of biofuels: The intersection of environmental, energy and agricultural policy. *Applied Econonomic Perspectives and Policy.* **32(1): 4–32.**

Deininger, K. 2011. Challenges posed by the new wave of farmland investment. *The Journal of Peasant Studies*, 38(2): 217–247.

Deininger, K. & Byerlee, D. (with Lindsay, J., Norton, A., Selod, H. & Stickler, M.) 2011. *Rising global interest in farmland. Can it yield sustainable and equitable benefits?* Washington, DC, World Bank.

Delgado, G.C. & Cardoso, J.S., eds. 2000. *A Universalização de Direitos Sociais no Brasil: A Previdência Rural nos Anos 90*. Brasília, IPEA (The Institute for Applied Economic Research).

Dercon, S. & Singh, A. 2012. *Investment in rural Ethiopia 1994–2009: a household perspective.* Background paper for *The State of Food and Agriculture 2012*. Rome, FAO.

Dercon, S., Gilligan, D.O., Hoddinott, J. & Woldehanna, T. 2009. The impact of agricultural extension and roads on poverty and consumption growth in fifteen Ethiopian villages. *American Journal of Agricultural Economic,* 91(4): 1007–1021.

Diakosavvas, D. 1990. Government expenditure on agriculture and agricultural performance in development countries: an empirical evaluation. *Journal of Agricultural Economics,* 41(3): 381–390.

Diao, X., Fan, S., Kanyarukiga S. & Yu, B. 2010. *Agricultural growth and investment options for poverty reduction in Rwanda.* IFPRI Research Monograph, Washington, DC, IFPRI.

Dias, P. 2012. *The determinants of household investment: a case-study exploring the relationship between access to credit and investment at the farm level in Nicaragua.* Background paper for *The State of Food and Agriculture 2012.* Rome, FAO.

Dillon, A., Sharma, M. & Zhang, X. 2008. *Nepal agriculture public expenditure review.* IFPRI paper prepared for the Department for International Development (DFID), London.

Dong, X.-Y. 2000. Public investment, social services and productivity of Chinese household farms. *Journal of Development Studies,* 36(3): 100–122.

Drayton, B. & Budinich, V. 2010. A new alliance for global change. *Harvard Business Review,* September (available at http://hbr.org/2010/09/a-new-alliance-for-global-change/ar/1).

Easterling, W.E., Aggarwal, P.K., Batima, P., Brander, K.M., Erda, L., Howden, S.M., Kirilenko, A., Morton, J., Soussana, J.-F., Schmidhuber, J. & Tubiello, F. 2007. Food, fibre and forest products. *In* M.L. Parry, O.F. Canziani, J.P. Palutikof, P.J. van der Linden & C.E. Hanson, eds. *Climate change 2007: impacts, adaptation and vulnerability. Contribution of Working Group II to the Fourth Assessment Report of the Intergovernmental Panel on Climate Change*, pp. 273–313. Cambridge, UK, Cambridge University Press.

Easterly, W. & Rebelo, S. 1993. Fiscal policy and economic growth: an empirical investigation. *Journal of Monetary Economics,* 32(2): 417–458.

Eastwood, R., Lipton, M. & Newell, A. 2010. Farm size. *In* P. Pingali and R. Evenson, eds. *Handbook of Agricultural Economics,* Vol. 4, pp. 3323–3397. Amsterdam, Elsevier.

Echeverría, R.G. & Beintema, N.M. 2009. *Mobilizing financial resources for agricultural research in developing countries: trends and mechanisms.* Global Forum for Agricultural Research (GFAR) Briefing Paper. Rome, GFAR.

Evenson, R.E. 2001. Economic impacts of agricultural research and extension. *In* B. Gardner and G. Rausser, eds. *Handbook of Agricultural Economics,* Vol. 1A, Chapter 11. Amsterdam, Elsevier.

Evenson, R.E. & Gollin, D. 2007. Contributions of national agricultural research systems to crop productivity. *In* R.E. Evenson, and P. Pingali, eds. *Handbook of agricultural economics*, Vol. 3, pp. 2420–2458. Amsterdam, Elsevier.

Evenson, R.E. & Fuglie, K.O. 2009. *Technology capital: the price of admission to the growth club.* Paper submission No 51398 at Conference for International Association of Agricultural Economists, Bejing, China, 16–22 August 2009.

Fan, S. & Rao, N. 2003. *Public spending in developing countries: Trends, determination, and impact.* EPTD Discussion Paper No. 99. Washington, DC, IFPRI.

Fan, S. & Saurkar, A. 2006. *Public spending in developing countries: trends, determination and impact* (mimeo).

Fan, S. & Zhang, X. 2008. Public expenditure, growth and poverty reduction in rural Uganda. *African Development Review,* 20(3): 466–496.

Fan, S., Gulati, A. & Thorat, S. 2008. Investment, subsidies, and pro-poor growth in rural India. *Agricultural Economics,* 39(2): 163–170.

Fan, S., Hazell, P. & Haque, T. 2000. Targeting public investments by agro-ecological zone to achieve growth and poverty alleviation goals in rural India. *Food Policy,* 25(4): 411–428.

Fan, S., Hazell, P. & Thorat, S. 2000. Government spending, agricultural growth and poverty in rural India. *American Journal of Agricultural Economics* 82(4): 1038–1051.

Fan, S., Yu, B. & Jitsuchon, S. 2008. Does allocation of public spending matter in poverty reduction? Evidence from Thailand. *Asian Economic Journal,* 22(4): 411–430.

Fan, S., Zhang, L. & Zhang, X. 2004. Reforms, investment and poverty in rural China. *Economic Development and Cultural Change,* **52(2): 395–421.**

FAO. 1947. *The State of Food and Agriculture 1947.* Rome.

FAO. 1949. *The State of Food and Agriculture 1949. A survey of world conditions and prospects.* Rome.

FAO. 2007. *The State of Food and Agriculture 2007. Paying farmers for environmental services.* Rome.

FAO. 2009a. *Food security and agricultural mitigation in developing countries: options for capturing synergies.* Rome.

FAO. 2009b. *The State of Food Insecurity in the World 2009. Economic crisis: impacts and lessons learned.* Rome.

FAO. 2010a. *"Climate-smart" agriculture: policies, practices and financing for food security, adaptation and mitigation.* Rome.

FAO. 2010b. Rural Income Generating Activities (RIGA) database (available at http://www.fao.org.economic/riga/en/).

FAO. 2011a. *Save and grow: a policy-maker's guide to the sustainable intensification of smallholder crop production.* Rome.

FAO. 2011b. *Food outlook: global market analysis,* November. Rome.

FAO. 2011c. Mapping actions for food security and nutrition (MAFSAN) web platform (available at www.mafsan.org).

FAO. 2011d. *The State of Food and Agriculture 2010–11. Women in agriculture: closing the gender gap.* Rome.

FAO. 2011e. *Land tenure and international investments in agriculture.* A report by the High Level Panel of Experts on Food Security and Nutrition of the Committee on World Food Security. Rome.

FAO. 2011f. *Report on expert meeting on international investment in the agriculture sector of developing countries.* Rome, Italy, 22–23 November.

FAO. 2011g. RAI Knowledge Platform: RAI Overview (available at http://www.responsibleagroinvestment.org/rai/node/232).

FAO. 2011h. *The State of the World's Land and Water Resources for Food and Agriculture. Managing systems at risk*, FAO Conference document C2011/32. Thirty-seventh Session. Rome, 25 June–2 July (available at: *www.fao.org/nr/solaw/solaw-home/en/*).

FAO. 2011i. *Global food losses and food waste, extent, causes and prevention*, by J. Gustavsson, C. Cederberg, U. Sonesson (Swedish Institute for Food, and Biotechnology) and R. van Otterdijk and A. Meybeck (FAO). Rome.

FAO. 2012a. FAOSTAT statistical database (available at faostat.fao.org).

FAO. 2012b. *Voluntary Guidelines on the Responsible Governance of Tenure of Land, Fisheries and Forests in the Context of National Food Security.* Rome.

FAO. 2012c. *Identifying opportunities for climate-smart agriculture investments in Africa*. (available at: http://www.fao.org/docrep/015/an112e/an112e00.pdf).

FAO. 2012d. *Trends and impacts of foreign Investment in developing country agriculture: evidence from case studies,* Rome.

FAO, IFAD & WFP. 2012. *The State of Food Insecurity in the World 2012. Economic growth is necessary but not sufficient to accelerate reduction of hunger and malnutrition.* Rome. FAO.

FAO, IFAD, UNCTAD & World Bank. 2012. *Principles for Responsible Agricultural Investment that Respects Rights, Livelihoods and Resources*. Synoptic version (available at http://www.fao.org/fileadmin/templates/est/INTERNATIONAL-TRADE/FDIs/RAI_Principles_Synoptic.pdf).

Fernandez, R. & Rodrik, D. 1991. Resistance to reform: status quo bias in the presence of individual-specific uncertainty. *American Economic Review,* 81(5): 1146–1155.

Ferroni, M. & Castle, P. 2011. Public-private partnerships and sustainable agricultural development. *Sustainability*, 2011(3):1064–1073.

FIAN (FoodFirst Information and Action Network). 2010. *Stop land grabbing now!* Online publication (available at http://www.fian.org/resources/documents/others/stop-land-grabbing-now/pdf).

Fischer, R.A., Byerlee, D. & Edmeades, G.O. 2009. *Can technology deliver on the yield challenge to 2050?* Paper prepared for the Expert

Meeting on How to Feed the World in 2050, organized by FAO, Rome, Italy, 24–26 June 2009.

Foster, A. & Rosenzweig, M. 2004. Agricultural productivity growth, rural economic diversity, and economic reforms: India, 1970–2000. *Economic Development and Cultural Change*, 52(3): 509–542.

Friis, C. & Reenberg, A. 2010. *Land grab in Africa: emerging land system drivers in a teleconnected world.* São José dos Campos, Brazil, GLP (The Global Land Project).

Fuglie, K.O. 2010. Sources of growth in Indonesian agriculture. *Journal of Productivity Analysis,* 33: 225–240.

Fuglie, K.O. 2012. Productivity growth and technology capital in the global agricultural economy. *In* K.O. Fuglie, S.L. Wang, & V.E. Ball, eds. *Productivity growth in agriculture: an international perspective.* Wallingford, UK, CAB International.

G8 (Group of Eight). 2009. *G8 Leaders Declaration: Responsible Leadership for a Sustainable Future.* Thirty-fifth G8 Summit, L'Aquila, Italy, 8–10 July 2009 (available at http://www.g8italia2009.it/static/G8_Allegato/G8_Declaration_08_07_09_final%2c0.pdf).

GEF (Global Environment Facility). 2012. *What is the GEF.* GEF website (available at http://www.thegef.org/gef/whatisgef).

Gertler, P., Martinez, S. & Rubio-Codina, M. 2012. Investing cash transfers to raise long-term living standards. *American Economic Journal: Applied Economics*, 4(1): 164–192.

Gilligan, D., Hoddinott J. & Taffesse, A. 2009. The impact of Ethiopia's productive safety net program and its linkages. *Journal of Development Studies*, 45(10): 1684–1706.

Gonzalez-Velosa, C. 2011. *The effects of emigration and remittances on agriculture: evidence from the Philippines.* Job market paper (available at ftp://ftp.cemfi.es/pdf/papers/wshop/JMP_Gonzalezvelosa_JAN.pdf).

Görgen, M., Rudloff, B., Simons, J., Üllenberg, A., Väth, S. & Wimmer, L. 2009. *Foreign direct investment (FDI) in land in developing countries.* Eschborn, Germany, Gesellschaft für Technische Zusammenarbeit (GTZ).

Government of Rwanda. 2009. *Agriculture Sector Investment Plan 2009–2012.* Rwanda, Ministry of Agriculture and Animal Resources.

Hallam, D. 2010. International investment in developing country agriculture: issues and challenges, *Agriregionieuropa*, No. 20, March.

Hayami, Y. & Ruttan, V.W. 1970. Agricultural productivity differences among countries. *American Economic Review,* 60(5): 895–911.

Hazell, P. & Haddad, L. 2001. *Agricultural research and poverty reduction.* Food, Agriculture and the Environment Discussion Paper 34. Washington, DC, IFPRI.

Herbel, D., Crowley, E., Ourabah Haddad, N. & Lee, M. 2012. *Good practices in building innovative rural institutions to increase food security.* Rome, FAO and IFAD.

Hoddinott, J. 2008. *Social safety nets and productivity enhancing investments in agriculture.* Paper presented at the conference "Convergence between Social Service Provision and Productivity Enhancing Investments in Development Strategies", Durban, South Africa, January 29–31. Washington, DC, IFPRI.

Hoff, K. & Stiglitz, J.E. 1997. Moneylenders and bankers: price-increasing subsidies in a monopolistically competitive market. *Journal of Development Economics*, 52(2): 429–462.

Horne, F. 2011. *Understanding land investment deals in Africa. Country report: Ethiopia.* Oakland, CA, USA, The Oakland Institute.

Huang, J. & Ma, H. 2010. *Capital formation and agriculture development in China*. Rome, FAO.

IFPRI (International Food Policy Research Institute). 2009. *Food security under stress from price volatility, agricultural neglect, climate change, and recession.* Presentation for the IPC Spring Seminar, Salzburg, Austria,11 May.

IFPRI. 2010. Statistics of Public Expenditure for Economic Development (SPEED). Online database (available at http://www.ifpri.org/book-39/ourwork/programs/priorities-public-investment/speed-database).

IFPRI. 2012a ASTI database (available at http://www.asti.cgiar.org/data/).

IFPRI. 2012b. Statistics of Public Expenditure for Economic Development (SPEED) database. Unpublished.

IMF (International Monetary Fund). 2001. *Government Finance Statistics Manual.* Washington, DC.

INTOSAI (International Organization of Supreme Audit Institutions) Working Group on Environmental Accounting. 2010. *Environmental accounting: current status and options for SAIs.* (also available at http://www.environmental-auditing.org/LinkClick.aspx?fileticket=s%2FFCvUzSKsk%3D&tabid=128&mid=568)

IPCC (Intergovernmental Panel on Climate Change). 2012. *Managing the risks of extreme* *events and disasters to advance climate change adaptation*. A Special Report of Working Groups I and II of the Intergovernmental Panel on Climate Change, edited by C.B. Field, V. Barros, T.F. Stocker, D. Qin, D.J. Dokken, K.L. Ebi, M.D. Mastrandrea, K.J. Mach, G.-K. Plattner, S.K. Allen, M. Tignor & P.M. Midgley. Cambridge, UK, and New York, USA, Cambridge University Press.

Jacoby, H.G. 2000. Access to markets and the benefits of rural roads. *The Economic Journal,* 110(465): 713–737.

Jha, R. 2007. *Investment and subsidies in Indian agriculture*. ASARC Working Paper 2007/03. Canberra, ACT, Australia, Australia South Asia Research Centre.

Kessides, C. 1993. *The contributions of infrastructure to economic development, a review of experience and policy implications*. World Bank Discussion Paper 213. Washington, DC, World Bank.

Key, N. & Roberts, M. J. 2007a. *Commodity payments, farm business survival, and farm size growth.* Economic Research Report No. 51. Washington, DC, United States Department of Agriculture (USDA).

Key, N. & Roberts, M.J. 2007b. Measures of trends in farm size tell differing stories. *AmberWaves,* 5(5):36–37.

Kolavalli, S., Birner, R., Benin, S., Horowitz, L., Babu, S., Asenso-Okyere, K., Thompson, N.M. & Poku, J. 2010. *Institutional and public expenditure review of Ghana's Ministry of Food and Agriculture*. IFPRI Discussion Paper 1020. Washington, DC, IFPRI.

Krueger, A., Schiff, M. & Valdés, A. 1988. Agricultural incentives in developing countries: measuring the effects of sectoral and economywide policies. *World Bank Economic Review,* 2(3): 255–272.

Krueger, A., Schiff, M. & Valdés, A. 1991. *The Political Economy of Agricultural Pricing Policy,* Vol. 1: Latin America, Vol. 2: Asia, and Vol. 3: Africa and the Mediterranean. Baltimore, USA, Johns Hopkins University Press for the World Bank.

Larson, D.F., Anderson J.R. & Varangis, P. 2004. Policies on managing risk in agricultural markets. *World Bank Research Observer,* 19(2):199–230.

Larson, D.F., Butzer, R., Mundlak, Y. & Crego, A. 2000. A cross-country database for sector investment and capital. *The World Bank Economic Review,* 14(2): 371–391.

Lee, D. 2011. *Accounting for natural resources and environmental goods and services in agricultural investment decisions: review and assessment.* Background paper prepared for *The State of Food and Agriculture 2012*. Rome, FAO.

Lin, J.Y. 1992. Rural reforms and agricultural growth in China. *American Economic Review,* 82(1): 34–51.

Lipper, L. & Neves, N. 2011. *Payments for environmental services: what role in sustainable agricultural development?* ESA working paper No. 11–20. Rome, FAO.

Lipton, M. 1977. *Why poor people stay poor: urban bias in world development.* Cambridge, Harvard University Press.

López, R. & Galinato, G.I. 2006. Should governments stop subsidies to private goods? Evidence from rural Latin America. *Journal of Public Economics,* 91 (2007): 1071–1094.

Lowder, S. & Carisma, B. 2011. *Financial resource flows to agriculture: a review of data on government spending, official development assistance and foreign direct investment.* ESA Working Paper No. 11–18, Rome, FAO.

Lowder, S., Carisma, B. & Skoet, J. 2012. *Who invests in agriculture and how much? An empirical review of the relative size of various investments in agriculture in low- and middle-income countries.* ESA Working Paper No. 12–09, Rome, FAO.

Lucas, R.E.B. 1987. Emigration to South Africa's mines. *American Economic Review,* 77(3): 313–330.

Government of Malawi, Ministry of Agriculture, Irrigation and Water Development. 2012. *Approaches to the implementation of conservation* agriculture *among promoters in Malawi. Baseline study (*available at http://www.moafsmw.org/ocean/docs/Research/Approaches%20to%20the%20Implemetation%20of%20CA%20among%20Promoters%20in%20Malawi-FINAL%208%20May%202012.pdf).

Maluccio, J. 2005. *Coping with the "coffee crisis" in Central America. The role of the Nicaraguan Red de Protección Social.* Discussion Paper 188. Washington, DC, IFPRI.

Maluccio, J. 2010. The impact of conditional cash transfers on consumption and investment in Nicaragua. *The Journal of Development Studies,* 46(1): 14–38.

Mansuri, G. 2007. **Credit layering in informal financial markets.** *Journal of Development Economics,* 84(2):715–730.

McCarthy, N., Lipper, L. & Branca, G. 2011. *Climate-smart agriculture: smallholder adoption and implications for climate change adaptation and mitigation.* Mitigation in Agriculture Series No. 4. Rome, FAO.

McMillan, J., Whalley J. & Zhu, L. 1989. The impact of China's economic reforms on agricultural productivity growth. *Journal of Political Economy*, 97(4): 781–807.

McNellis, P.E. 2009. *Foreign direct investments in developing country agriculture: the emerging role of private sector finance.* FAO Commodity and Trade Policy Research Working Paper No. 28. Rome, FAO.

Mendola, M. 2008. Migration and technological change in rural households: complements or substitutes? *Journal of Development Economics*, 85(1–2): 150–175.

Meyer, Richard L. 2011. *Subsidies as an instrument in agricultural development finance: review.* Joint Discussion Paper of the Joint Donor CABFIN Initiative. Washington, DC, World Bank.

Millennium Ecosystem Assessment. 2005. *Ecosystems and human well-being: synthesis.* Washington, DC, Island Press.

Miller, C. & Jones, L. 2010. *Agricultural value chain finance: tools and lessons.* Rome, FAO and Rugby, UK, Practical Action Publishing.

Miller, C., Richter, S., McNellis, P. & Mhlanga, N. 2010. *Agricultural investment funds for developing countries.* Rome, FAO.

Miluka, J., Carletto, G., Davis, B. & Zezza, A. 2007. *The vanishing farms? The impact of international migration on Albanian family farming.* ESA Working Paper No. 07-09. Rome, FAO.

Mogues, T. 2011. **The bang for the birr: public expenditures and rural welfare in Ethiopia.** *Journal of Development Studies*, 47(5): 735–752.

Mogues, T. 2012. What determines public expenditure allocations? A review of theories, and implications for agricultural public investments. ESA Working Paper No. 12-06, Rome, FAO.

Mogues, T., Yu, B., Fan, S. & L. McBride. 2012 (forthcoming). *The impacts of public investments in and for agriculture: synthesis of the existing evidence.* ESA Working Paper No. 12-07, Rome, FAO.

Møller, V. & Ferreira, M. 2003. *Getting by... benefits of non-contributory pension income for older South African households.* University of Cape Town, South Africa, Institute for Ageing in Africa. (mimeo)

Morris, J., Williams, A.G. & Audsley, E. 2007. Greening the lilies – environmental accounting for agriculture. *Journal of the Royal Agricultural Society of England,* 168: 1–10.

Mu, R. & van de Walle, D. 2007. *Rural roads and local market development in Vietnam.* Policy Research Working Paper 4340. Washington, DC, World Bank.

Mundlak, Y., Larson, D. F. & Butzer, R. 2004. *The determinants of agricultural production: a cross-country analysis.* Policy Research Working paper 1827. Washington, DC, World Bank.

Nagayets, O. 2005. *Small farms: current status and key trends.* Information brief prepared for the Future of Small farms Research Workshop, Wye College, Kent, UK, 26–29 June 2005.

Nelson, G.C., Rosegrant, M.W., Koo, J., Robertson, R., Sulser, T., Zhu, T., Ringler, C., Msangi, S., Palazzo, A., Batka, M., Magalhaes, M.,Valmonte-Santos, R., Ewing, M. & Lee, D. 2009. *Climate change: impacts on agriculture and costs of adaptation.* Washington, DC, IFPRI.

NEPAD (The New Partnership for Africa's Development). 2010a. Global Donor Platform NEPAD pillar documents (available at http://www.nepad-caadp.net).

NEPAD. 2010b. *Implementing CAADP for Africa's Food Security needs: a progress report on selected activities.* Midrand, South Africa (also available at http://www.nepad-caadp.net/pdf/Final%20CAADP%20MAF%20Validation%20Report.pdf).

NEPAD. 2010c. *The Comprehensive Africa Agriculture Development Programme (CAADP) in practice: highlighting the successes.* Midrand, South Africa (also available at http://www.nepad-caadp.net/pdf/Highlighting%20the%20successes%20280611%20v3%200%20web.pdf).

Oberai, A. & Singh, H.K.M. 1983. *Causes and consequences of internal migration.* Delhi, Oxford University Press.

OECD (Organisation for Economic Co-operation and Development). 2010. Working Group on Economic Aspects of Biodiversity (WGEAB). *OECD Expert Workshop on Enhancing the Cost-effectiveness of Payment for Ecosystem Services (PES) Summary record.* Paris.

OECD. 2011. *Policy framework for investment in agriculture: policy guidance for promoting private investment in agriculture in Africa.* Preliminary version. Paris.

OECD. 2012a. Credit Reporting System Aid Activities (CRS) database (available at http://stats.oecd.org/Index.aspx?datasetcode = CRS1).

OECD. 2012b. *Environmental Outlook to 2050.* OECD, Paris.

OECD-FAO. 2012. *OECD-FAO Agricultural Outlook: 2012–2021.* Paris, OECD and Rome, FAO.

Olson, M. 1965. *The logic of collective action.* New Haven, USA, Yale University Press.

Olson, M. 1985. Space, agriculture and organisation. *American Journal of Agricultural Economics*, 67(5): 928–937.

Omuru, E. & **Kingwell, R.** 2006. Funding and managing agricultural research in a developing country: a Papua New Guinea case study. *International Journal of Social Economics,* 33(4): 316–330.

Otsuka, K., Estudillo, J.P. & Sawada, Y. 2009. *Rural poverty and income dynamics in Asia and Africa*. London, Routledge.

Palmer-Jones, R. & Sen, K. 2003. What has luck got to do with it? A regional analysis of poverty and agricultural growth in rural India. *Journal of Development Studies,* 40(1): 1–31.

Pray, C.E., Fuglie, K.O. & Johnson, D. 2007. Private agricultural research. *In* R.E. Evenson & P. Pingali, eds. *Handbook of agricultural economics*, Vol. 3, pp. 2605–2633. Amsterdam, Elsevier.

Pretty, J.N., Noble, D., Bossio, J., Dixon, R.E., Hine, F.W., Penning de Vries, T. & Morison, J.I.L. 2006. Resource-conserving agriculture increases yields in developing countries. *Environmental Science and Technology*, 40: 4.

Pritchett, L. 1996. *Mind your P's and Q's. The cost of public investment is not the value of public capital.* World Bank Policy Research Working Paper 1660. Washington, DC, World Bank.

Rajkumar, A.S. & Swaroop, V. 2008. Public spending and outcomes: does governance matter? *Journal of Development Economics*, 86: 96–111.

Rapsomanikis, G. & Vezzani, A. 2012 (forthcoming). *Lagging behind. An investigation on the dynamics of agricultural labour productivity*. ESA working paper series. Rome, FAO.

Ravallion, M. & Chen, S. 2004. *China's (uneven) progress against poverty*. World Bank Policy Research Working Paper 3408. Washington, DC, World Bank.

ReSAKSS (Regional Strategic Analysis and Knowledge Support System). 2011. Africa wide CAADP targets data (available at http://www.resakss.org/).

Resnick, D. & Birner, R. 2006. *Does good governance contribute to pro-poor growth? A review of the evidence from cross-country studies.* DSDG Discussion Paper No. 30. Washington, DC, IFPRI.

Rosegrant, M., Kasryno, F. & Perez, N.D. 1998. Output response to prices and public investment in agriculture: Indonesian food crops. *Perez Journal of Development Economics,* 55(2): 333–352.

Ruben, R. & Pender, J. 2004. Rural diversity and heterogeneity in less-favoured areas: the quest for policy targeting. *Food Policy*, 29(4): 303–320.

Sabates-Wheeler, R. & Devereux S. 2010. Cash transfers and high food prices: explaining outcomes on Ethiopia's productive safety net programme. *Food Policy*, 35(4): 274–285.

Schiff, M. & Valdés, A. 2002. Agriculture and the macroeconomy, with emphasis on developing countries. *In* B. Gardner & G. Rausser, eds. *Handbook of Agricultural Economics*, Vol. 2A, pp. 1421–1454. Amsterdam, Elsevier.

Schmidhuber, J. & Bruinsma, J. 2011. Investing towards a world free of hunger: lowering vulnerability and enhancing resilience. *In* A. Prakash, ed. *Safeguarding food security in volatile global markets*. Rome, FAO.

Schmidhuber, J., Bruinsma, J. & Boedeker, G. 2009. *Capital requirements for agriculture in developing countries to 2050.* Paper presented at the FAO Expert Meeting on "How to Feed the World in 2050", Rome, FAO, 24–26 June 2009.

Schwarzer, H. 2000. *Impactos socio-econômicos do sistema de aposentadorias rurais no Brasil: evidências empíricas de um estado de caso no estado do pará.* Institute for Applied Economic Research (IPEA) Texto para Discussão 729. Rio de Janeiro, Brazil, IPEA.

Sen, B. 2003. Drivers of escape and descent: changing household fortunes in rural Bangladesh. *World Development*, 31(3): 513–534.

Short, C., Barreiro-Hurlé, J. & Balié, J. 2012. *Analysis of price incentives and disincentives for maize in 10 African countries.* MAFAP Technical Notes. Rome, FAO.

Skees, J. R. 2008. Challenges for use of index-based weather insurance in lower income countries. *Agricultural Finance Review,* 68(1): 197–217.

Spielman, D.J., Hartwich, F. & von Grebmer, K. 2007. *Sharing science, building bridges, and enhancing impact: public-private partnerships in the CGIAR.* IFPRI Discussion Paper 708. Washington, DC, IFPRI.

Suphannachart, W. & Warr, P. 2011. Research and productivity in Thai agriculture. *Australian*

Journal of Agricultural and Resource Economics, 55(1): 35–52.

Suryahadi, A., Suryadarma, D. & Sumarto, S. 2009. The effects of location and sectoral components of economic growth on poverty: evidence from Indonesia. *Journal of Development Economics*, 89(1): 109–117.

Thorbecke, E. & Jung, H.-S. 1996. A multiplier decomposition method to analyse poverty alleviation. *Journal of Development Economics*, 48(2): 279–300.

Transnational Institute. 2011. *It is time to outlaw landgrabbing, not to make it "responsible"!* (available at http://www.tni.org/sites/www.tni.org/files/RAI-EN-1.pdf).

Tsegai, D. 2004. *Effects of migration on the source communities in the Volta Basin of Ghana: potential links of migration, remittances, farm and non-farm self-employment activities.* Economics and Technological Change Working Paper. Bonn, Germany, University of Bonn.

UNCTAD (United Nations Conference on Trade and Development). 2011. Internal data not publically available at the country level.

van der Mensbrugghe, D. 2005. *Linkage technical reference document, version 6.0.* Prepared by Development Prospects Group (DECPG). Washington, DC, World Bank.

Vermeulen, S. & Cotula, L. 2010. *Making the most of agricultural investment: a survey of business models that provide opportunities for smallholders.* Geneva, Switzerland, FAO and IIED.

Visser, O. & Spoor, M. 2011. Land grabbing in the post-Soviet region [in Russian]. *In* A. Nikulin & T. Shanin, eds. *Krestyanovedenie: the study of peasantry*. Moscow, Rospen / MSSES.

von Braun, J. & Meinzen-Dick, R.S. 2009. *"Land grabbing" by foreign investors in developing countries: risks and opportunities.* Policy Brief 13. Washington, DC, IFPRI.

von Braun, J., Gulati, A. & Fan, S. 2005. *Agricultural and economic development strategies and the transformation of China and India*. Washington, DC, IFPRI.

Vorley, B. & Proctor, F. 2008. *Inclusive business in agrifood markets: evidence and action.* A report based on proceedings of an international conference held in Beijing, China, 5–6 March (available at http://www.regoverningmarkets.org/en/filemanager/active?fid=).

Wall Street Journal. 2010. *Private sector interest grows in African farming* (available at http://online.wsj.com/article/SB10001424052702303467004575574152965709226.html).

Warner, M., Kahan, D. & Lehel, S. 2008. *Market-oriented agricultural infrastructure: appraisal of public-private partnerships.* Agricultural Management, Marketing and Finance Occasional Paper 23. Rome, FAO.

Wiggins, S. & Brooks, J. 2010. *The use of input subsidies in developing countries*. Paper presented at the Global Forum on Agriculture, OECD, Paris, 29–30 November, 2010.

World Bank. 2004. *World Development Report 2005. A better investment climate for everyone*. Washington, DC.

World Bank. 2006a. *Where is the wealth of nations? Measuring capital for the 21st century.* Washington, DC.

World Bank. 2006b. *The rural investment climate: it differs and it matters.* Agriculture and Rural Development Department Report No. 36543-GLB. Washington, DC.

World Bank. 2007a. *World Development Report 2008. Agriculture for development*. Washington, DC.

World Bank. 2007b. *Philippines: agriculture public expenditure review.* Technical working paper 40493. Washington, DC.

World Bank. 2008. *Nigeria: agriculture public expenditure review.* Report No. 44000-NG. Washington, DC.

World Bank. 2010a. *Uganda: agriculture public expenditure review*. Report No. 53704-UG. Washington, DC, World Bank.

World Bank. 2010b. *World Development Indicators* (available at http://data.worldbank.org/data-catalog/world-development-indicators/wdi-2010; accessed 5 July 2011).

World Bank. 2011a. *Mozambique: analysis of public expenditure in agriculture*. World Bank Report No. 59918-MZ. Washington, DC.

World Bank. 2011b. *United Republic of Tanzania public expenditure review*. Report No. 64585-TZ. Washington, DC.

World Bank. 2011c. *World Bank Governance Indicators* (available at http://databank.worldbank.org).

World Bank. 2011d. Doing business database (available at http://www.doingbusiness.org/data).

World Bank. 2011e. *Practitioners toolkit for agriculture public expenditure analysis.* Washington, DC/World Bank and UK Department for International Development.

World Bank. 2012. *World Development Indicators* (available at http://databank.worldbank.org).

Wunder, S., Engel, S. & Pagiola, S. 2008. Taking stock: a comparative analysis of payments for environmental services programs in developed and developing countries. *Ecological Economics*, 65(4): 834–852.

Xu, Z., Burke, W.J., Jayne T.S. & Govereh, J. 2009. Do input subsidy programs 'crowd in' or 'crowd out' commercial market development? Modeling fertilizer demand in a two-channel marketing system. *Agricultural Economics*, 40(1): 79–94.

Zhang, X. 2004. *Security is like oxygen: evidence from Uganda*. DSDG Discussion Paper No. 6. Washington, DC, IFPRI.

Zhang, X., Fan, S., Zhang, L. & Huang, J. 2004. Local governance and public goods provision in rural China. *Journal of Public Economics*, 88(12): 2857–2851.

Zimmermann, R., Bruntrüp, M., Kolavalli, S. & Flaherty, K. 2009. *Agricultural policies in sub-Saharan Africa: understanding CAADP and APRM policy processes*. Study 48. Bonn, Germany, The German Development Institute (DEI).

Special chapters of *The State of Food and Agriculture*

In addition to the usual review of the recent world food and agricultural situation, each issue of this report since 1957 has included one or more special studies on problems of longer-term interest. Special chapters in earlier issues have covered the following subjects:

1957	Factors influencing the trend of food consumption
	Postwar changes in some institutional factors affecting agriculture
1958	Food and agricultural developments in Africa south of the Sahara
	The growth of forest industries and their impact on the world's forests
1959	Agricultural incomes and levels of living in countries at different stages of economic development
	Some general problems of agricultural development in less-developed countries in the light of postwar experience
1960	Programming for agricultural development
1961	Land reform and institutional change
	Agricultural extension, education and research in Africa, Asia and Latin America
1962	The role of forest industries in the attack on economic underdevelopment
	The livestock industry in less-developed countries
1963	Basic factors affecting the growth of productivity in agriculture
	Fertilizer use: spearhead of agricultural development
1964	Protein nutrition: needs and prospects
	Synthetics and their effects on agricultural trade
1966	Agriculture and industrialization
	Rice in the world food economy
1967	Incentives and disincentives for farmers in developing countries
	The management of fishery resources
1968	Raising agricultural productivity in developing countries through technological improvement
	Improved storage and its contribution to world food supplies
1969	Agricultural marketing improvement programmes: some lessons from recent experience
	Modernizing institutions to promote forestry development
1970	Agriculture at the threshold of the Second Development Decade
1971	Water pollution and its effects on living aquatic resources and fisheries
1972	Education and training for development
	Accelerating agricultural research in the developing countries
1973	Agricultural employment in developing countries
1974	Population, food supply and agricultural development
1975	The Second United Nations Development Decade: mid-term review and appraisal
1976	Energy and agriculture
1977	The state of natural resources and the human environment for food and agriculture
1978	Problems and strategies in developing regions
1979	Forestry and rural development
1980	Marine fisheries in the new era of national jurisdiction
1981	Rural poverty in developing countries and means of poverty alleviation
1982	Livestock production: a world perspective
1983	Women in developing agriculture
1984	Urbanization, agriculture and food systems

1985	Energy use in agricultural production
	Environmental trends in food and agriculture
	Agricultural marketing and development
1986	Financing agricultural development
1987–88	Changing priorities for agricultural science and technology in developing countries
1989	Sustainable development and natural resource management
1990	Structural adjustment and agriculture
1991	Agricultural policies and issues: lessons from the 1980s and prospects for the 1990s
1992	Marine fisheries and the law of the sea: a decade of change
1993	Water policies and agriculture
1994	Forest development and policy dilemmas
1995	Agricultural trade: entering a new era?
1996	Food security: some macroeconomic dimensions
1997	The agroprocessing industry and economic development
1998	Rural non-farm income in developing countries
2000	World food and agriculture: lessons from the past 50 years
2001	Economic impacts of transboundary plant pests and animal diseases
2002	Agriculture and global public goods ten years after the Earth Summit
2003–04	Agricultural biotechnology: meeting the needs of the poor?
2005	Agriculture trade and poverty: can trade work for the poor?
2006	Food aid for food security?
2007	Paying farmers for environmental services
2008	Biofuels: prospects, risks and opportunities
2009	Livestock in the balance
2010–11	Women in agriculture: closing the gender gap for development